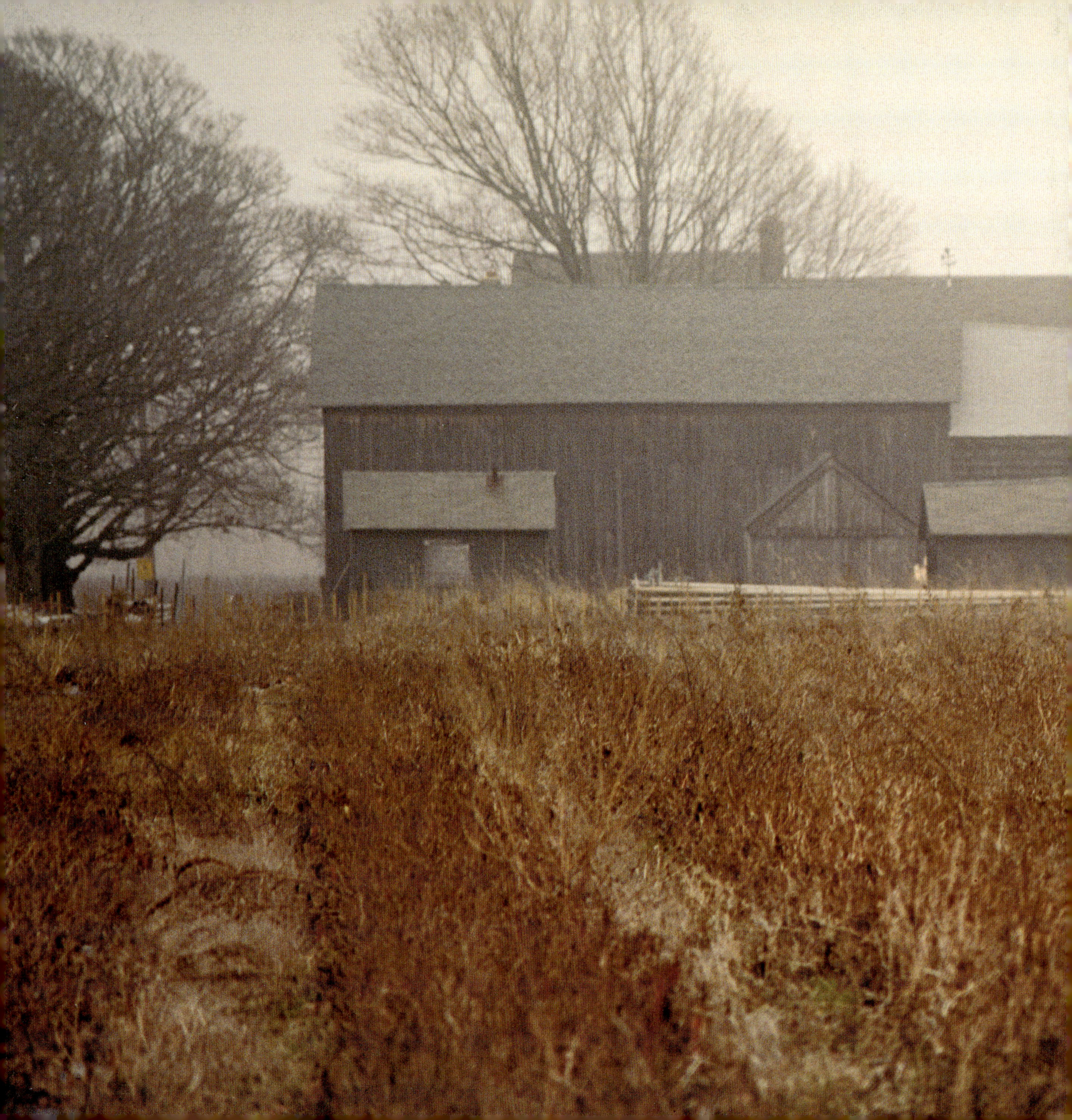

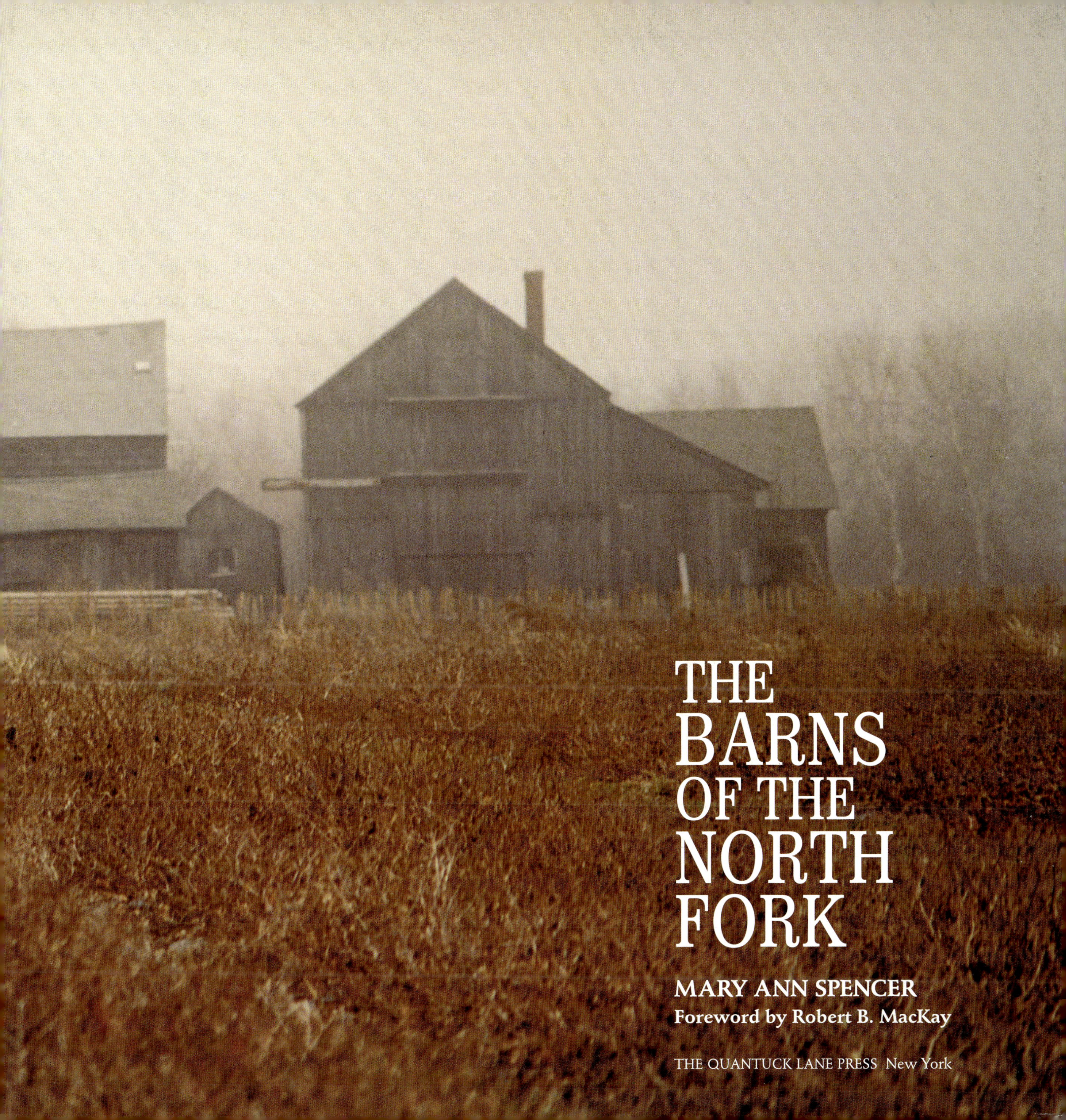

THE BARNS OF THE NORTH FORK

MARY ANN SPENCER

Foreword by Robert B. MacKay

THE QUANTUCK LANE PRESS New York

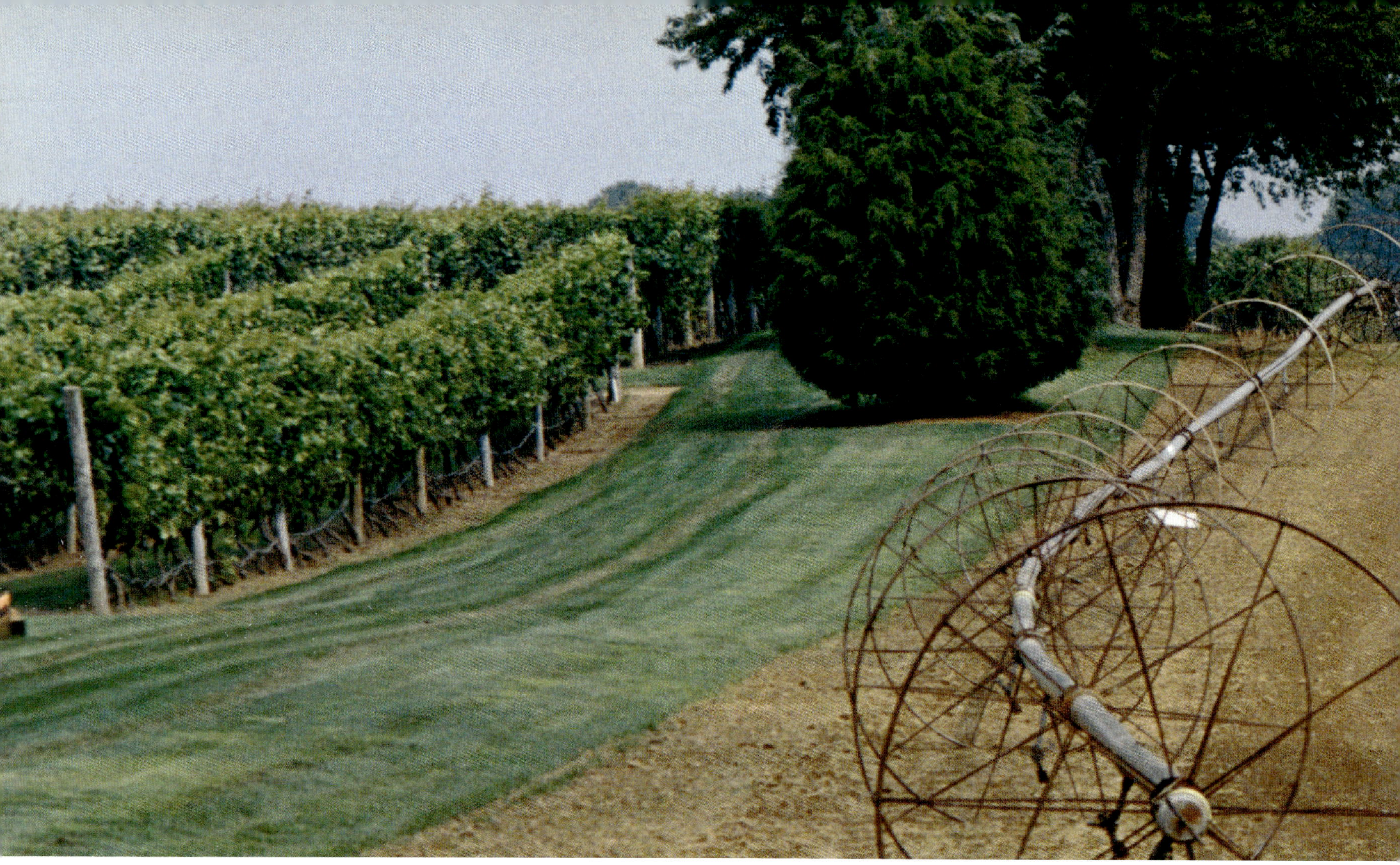

The Barns of the North Fork
Mary Ann Spencer

Printed in Italy
First Edition

Library of Congress Cataloging-in-Publication Data
Spencer, Mary Ann.
Barns of the North Fork / Mary Ann Spencer ; introduction by Robert B. MacKay.
p. cm.
Includes bibliographical references and indexes.
ISBN 1-59372-014-9
1. Barns—New York (State)—Long Island—Catalogs. I. Title.
NA8230.S67 2005
728'.922'0974721—dc22 2005006759

The text of this book is composed in Stempel Schneidler
with the display in Digi Antiqua.
Book design and composition by Laura Lindgren
Manufacturing by Mondadori Printing, Verona

The Quantuck Lane Press
www.quantucklanepress.com
Distributed by W. W. Norton & Company
500 Fifth Avenue
New York, NY 10110
www.norton.com

W. W. Norton & Company Ltd.
Castle House
75/76 Wells Street
London WIT 3QT

1 2 3 4 5 6 7 8 9 0

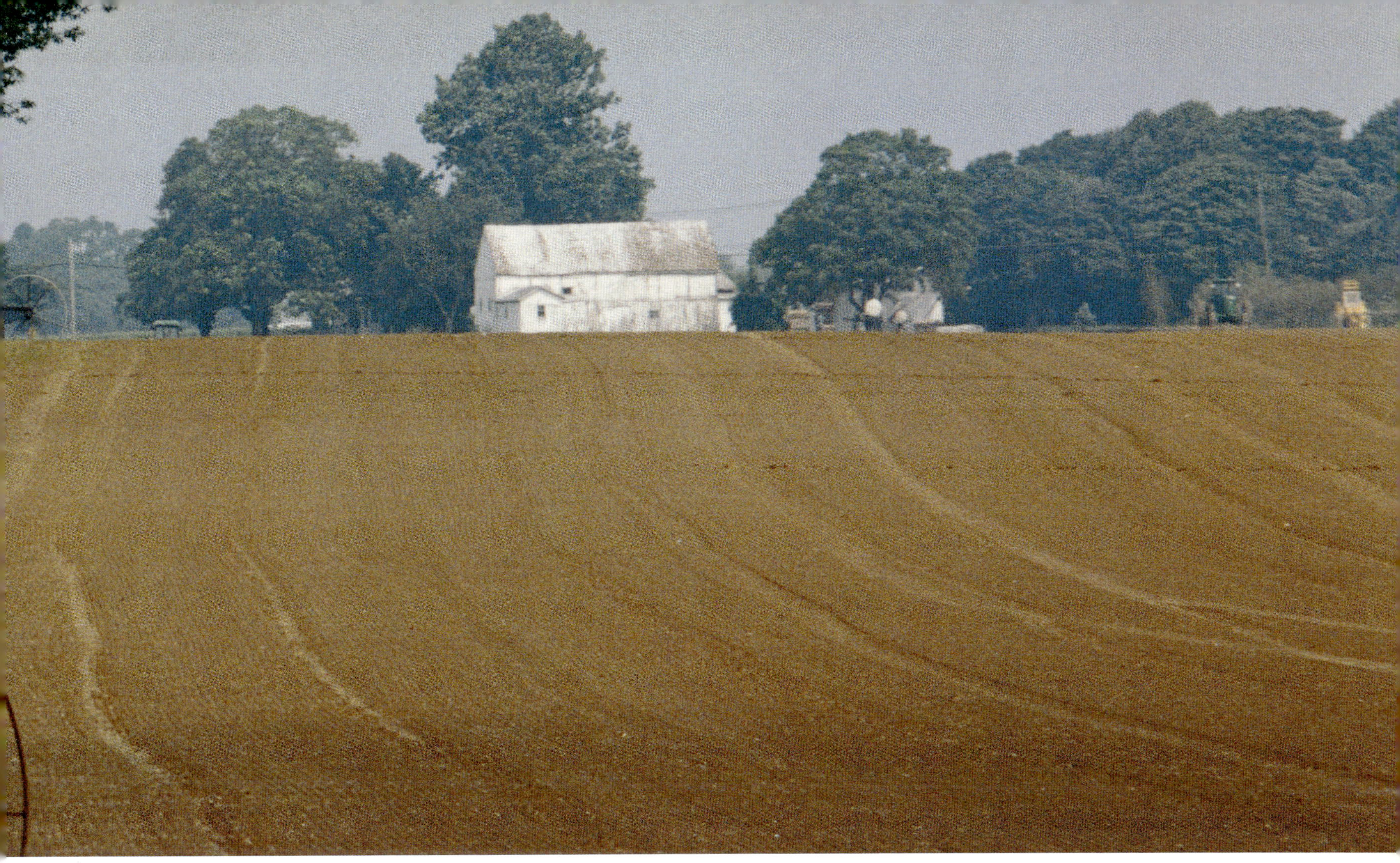

Contents

Foreword

Barns were the defining features of the rural landscape on Long Island for more than two and a half centuries. These majestic structures dotted the agrarian landscape from the city line to Montauk. Handsome and extensive, barns are seen, for example, throughout the work of Edward Lange (1846–1912), the German émigré artist, who compiled a remarkable portrait of Long Island in the 1870s and 1880s. The condition of its barns and outbuildings was an important factor in establishing the value of a farm before the turn of the nineteenth century. However, with the advent of the automobile age, the farms that had not been assembled to form the great estates soon made way for Levittown and post–World War II developments in Nassau and much of Suffolk counties. Today, these once omnipresent buildings have largely disappeared from most of our communities. The greatest concentration survives, not surprisingly, where farming still thrives on Long Island's North Fork, where these eminently adaptable buildings continue to meet the needs of farmers along Sound Avenue and Route 25.

Photographer Mary Ann Spencer, a native of Wyoming, has a particular awareness of this building type. Barns punctuated the wide-open landscape of the plains and prairies of her youth. For more than a quarter of a century, Mary Ann has been recording North Fork barns through every season and at every hour and has compiled a remarkable portrait of these quickly disappearing features of our agrarian past. She has inventoried, recorded, and now celebrates these architectural resources. In doing so, she is opening our eyes to an architecture worth saving and to a landscape that will never be the same.

Robert B. MacKay, Ph.D., Director,
Society for the Preservation of Long Island Antiquities

Introduction

The landscapes that I remember from my childhood are the sagebrush prairies that spread across Wyoming and the endless plains that begin at the edge of the Black Hills of South Dakota. My eye is drawn to evidence of settlement—barbed-wire fencing and windmills, singular schoolhouses and isolated homesteads. When a ranch or farm appears in this sparse setting the barn announces it. And when a homestead has been abandoned the barn remains long after the house is gone.

Perhaps that is why I visually select barns in any setting. I certainly find them pleasing. I like their scale and their simplicity, their variety and individuality. I like that they signify a presence or a history of agriculture, and that they appear unexpectedly in urban areas. And I like the fact that most barn owners like barns, too.

This affinity for barns led to a desire to do an inventory of the barns of Suffolk County, the eastern portion of Long Island. Suffolk County has an area of 911 square miles. Settled in 1640, it is one of the oldest counties in the country. For much of the nineteenth century the eastern section was significantly more populous than the western portion. Farming and fishing were at the region's core.

The county's agricultural history remains central to the economy and to the culture of the North Fork, and barns abound. When English settlers arrived on the glacial moraines that define Long Island they found that the land between the rocky deposits was essentially flat, virtually free of stones, and covered with fertile topsoil. The Indians had cleared large portions by burning and so the land was ready for planting.

On the North Fork English settlers built barns of the form familiar to them, the three-bay English hay barn. Allen Noble describes this barn in *Wood, Brick & Stone* as "a small rectangular barn. Although most widely identified as an English barn, the structure has been variously termed a New England, Connecticut, Yankee, and three-bay or two-bay barn, the last from its basic plan, which is that of a central floor area or runway with two spaces of roughly equal size on either side. Thus, they are two-bays if the runway is not counted and three-bays if it is. Above the bays is a loft for hay." These barns were well suited to the needs of subsistence farming. Once a pattern for barn building has been established in a given area of the country it is common for it to persist over time; the style of the English hay barn continued in popularity on Long Island through the nineteenth century.

Nine hundred eleven square miles is a large chunk of land so I decided to limit the initial survey to the North Fork and, further, to the Town of Southold. The Town of Southold is roughly fifty-eight square miles (including the Incorporated Village of Greenport). I secured a grant through the Old House Society in Cutchogue and began work in the fall of 2001.

There is a time for everything. Barn survey work is best done after the leaves have fallen because the barns are easier to locate and much easier to photograph. And it helps to wait even a bit longer. I began on the eastern tip of the island in Orient Point on an October morning. I stood watching Danny Latham drive toward me across a field on his tractor, and he remained seated on his idling tractor while I explained that I was interested in surveying his barn. "That would be fine," he said, "but could you come back after I finish the harvest?"

The people of Southold were welcoming and responsive to the project. When I had completed the survey of a community I asked for assistance from local residents—I showed them what I had found and asked, "What have I included that should be excluded? What have I missed?" Bob and Lillian White in Greenport joined in the spirit of the hunt; they invited me in for coffee and lovingly and knowledgeably illuminated the history of their area. Then we hopped in the car and they showed me several that I had missed because the barns were tucked away out of sight or had been converted to homes.

The purpose of the inventory was to do a reconnaissance-level survey of all the barns. Within the scope of the survey I attempted to locate, describe, photograph, and, whenever possible, document the history of each barn. Barns are functional agricultural structures whose names often indicate their particular use, e.g., hay barn, corncrib, carriage house, dairy barn. I didn't "count" attendant structures such as corncribs or chicken coops or farm sheds as barns, but they were noted. I did include carriage houses, which have been described as "urban barns" and were built to stable horses and shelter carriages. Since a barn could be used for these purposes few carriages houses appear on farms.

Barn construction continues on the North Fork. Most of these new barns are designed and built for agricultural use, and some are built as stables. Many but certainly not all of the newer barns are metal pole barns, sometimes known as "Butler" barns or "Morton" barns. The term "pole barn" came from the early practice of setting four poles in the ground with an attached roof providing shelter for crops or livestock; in time, sides were added to the poles. Metal barns haven't the aesthetic appeal of timber-framed wood-sided barns but they are barns all the same and they are included in the survey.

Barns are visual icons in the rural landscape, but more importantly they chronicle the evolution of farming in a particular region. Farm buildings need to be functional and therefore they are reconfigured, converted, and adapted in order to meet changing agricultural needs. I was talking with Bill Lindsay about his farm in Mattituck, which was begun in 1890, and about the history of his barns. He explained that his father gathered materials for one of the barns from the New York World's Fair in the 1930s. Bill indicated a building that appeared to be a shed and explained that it was in fact a resided corncrib.

Barns are often expanded by adding wings, the most common of which is the shed addition. The need for additional space often entails building a contiguous barn in which the second barn commonly shares one wall with the first. In some instances two such barns were built within a few years of one another, but it is not uncommon to find adjoining barns built a century apart. Hallock Tuthill's double barn on Alvah's lane contains an eighteenth-century three-bay English barn alongside an early-twentieth-century barn of roughly the same size. Hallock smiles when he reminds you that his first and last names are both Long Island farming families of long standing.

Older barns have been elevated and placed onto newer structures in order to accommodate farm machinery. When barns were lost to hurricanes or fire, very often a replacement barn was built over the remaining foundation or cellar. In the past twenty years vineyards represent an expanding agricultural business, and the preexisting barns on lands purchased by vineyards have been renovated to accommodate wine processing and to serve as tasting rooms. On the North Fork a great many barns have been converted for potato storage, which involves buttressing the walls of the barn by means of cinder blocks or cement or soil to support the weight of the crop.

Both in the building and in the expansion of barns timber framers recycle portions of frames. Eric Keil showed me one of his farm structures that has framing elements dating back to the seventeenth century. Dan and Prudence Heston have an English-style barn in Cutchogue that was built in 1852, and incorporated in the barn's frame are timbers from the 1732 Cutchogue Presbyterian church, which was dismantled in 1850. White oak knee braces flank the center bay, the east wall contains a reused wall girt and a section of a summer beam, and there are notched and pegged roof rafters used elsewhere. Prudence Heston's maiden name is Wickham—the Wickhams have farmed in Cutchogue for generations.

Some barns were built for a use that has not survived. Carriage houses are a clear example; they remain scattered throughout the North Fork and vary from small, unpretentious structures to elaborate estate outbuildings. Another forgone use is that of the baymen's shacks, or cottages, in East Marion. This was a fishing community and most of the land is original forest because it was never cleared for farming. The baymen's shacks resemble small carriage houses and that was what I thought they were. However, when I went back through the community with Gordon Racket he explained that they were built and used by fishermen for tarring and mending their nets.

When barns are no longer used for agricultural purposes—when they are abandoned in the landscape or the land surrounding them is sold for development—they are either lost to neglect or moved, adapted, or reused. Many such North Fork barns have been converted to workshops, warehouses, stores, office spaces, and more than a few have become homes. These renovations span the gamut, and many retain aspects of the barns' history, their interest, and their beauty—but the barns have lost their context.

Historic preservation begins with an accounting. More than seven hundred barns were surveyed. Businesses and homes constitute 6 percent of the survey. Townwide, 43 percent of the barns remain in agricultural use. In the farming communities the percentage increases to 52 percent. Only a handful of the barns that I found are collapsing from neglect, but each abandoned deteriorating barn is a strong visual statement.

In 1970–71 Charles Tichy began a survey of pre-1900 Suffolk County barns still in a farm setting. His survey (aided by a grant from the New York State Council on the Arts to the Nassau County Historical Museum) was not finished, but the survey forms that were completed include histories and diagrams of framing. Using his field notes, one of the questions that I tried to answer was how many of the barns noted in the 1970–71 survey survive. Approximately 15 percent of the barns surveyed by Charles Tichy are gone.

Statistics and photographs outline and illustrate the project but they cannot do justice to the wonderful experience of meeting the people who own and care for these barns. In Southold Helen Krukowski reminisced about her father, Felix, and his love of farming; because these structures remind her of him she reroofed the main barn. Down the road Cris Baiz has planted a vineyard on his family farm. The land slopes gently to the sea. His grandfather, the last farmer before him, raised chickens. The barn, circa 1860, rests on the original stone foundation and was extended for use as a milking barn in 1921. Cris hopes to restore the coops and raise chickens again.

I rode to work in the mornings along Sound Avenue past farms in Riverhead. These barns are not included in the survey itself but every now and then the light would be right and I couldn't resist photographing them, and some of these photographs are included in this book. On the way home I usually went a bit out of the way in order to amble along Oregon Road, which parallels Route 48. It begins at the end of Bridge Lane in Cutchogue and runs west into East Mattituck through Oregon. Oregon is the heartland. Farms stretch to the horizon uninterrupted by stores or parking lots. There are potato barns symbolizing the last great Long Island crop and there are vineyards and sod farms and nurseries representing new agricultural endeavors. You won't find Oregon on all of the maps but the people who live on the North Fork know and cherish it. It was the best part of my day.

Allen G. Noble, *Wood, Brick & Stone: The North American Settlement Landscape, Volume 2: Barns and Farm Structures,* 1984, The University of Massachusetts Press.

Acknowledgments

When I finished surveying a community I went back and asked residents to go over the material with me. I am deeply grateful to (starting on the east end and working west): Courtney Burns and Alan Bull from the Oysterponds Historical Society, Gordon Racket from East Marion, Bob and Lillian White from Greenport, Jim Rich who covered both Peconic and Southold, Russell Thomes from the Southold Historical Society, Vincent and Helen P. Krupski from Oregon Road, John Koroleski and Myron Young from Mattituck, and Norman Wamback from the Mattituck Historical Society.

The Old House Society in Cutchogue sponsored the survey. Jim Grathwohl, president of the Old House Society, helped in Cutchogue and with contacts in each community. He gave unselfishly of his time and his enthusiasm.

Photographic printing for the book was supported by a 2004 Independent Projects award from the Architecture, Planning, and Design Program of the New York State Council on the Arts.

I enticed Zach Studenroth, director of the Whaling Museum in Sag Harbor, to join me in the field from time to time. He taught me a great deal about barns.

Bob MacKay, director of the Society for the Preservation of Long Island Antiquities, saw the New York State Barn Coalition booth at the state fair one year and came back to report, "You're not the only one who loves barns!" Among Bob's many fine qualities is his unwavering encouragement for people involved in historic preservation. His support and guidance were invaluable throughout the project.

Jim Mairs, my editor and publisher, shares this love of barns. He is gracious, generous, and has a great eye. Working with him has been a joy.

My family is unfailingly there for me. Our son, David, has always cherished my work. Our daughter Danielle and her friend Phyllis Thompson Reid were incredibly helpful with their advice. Danielle's artistry and wisdom were indispensable. Finally, I wish to thank my husband, Joel, for his assistance, encouragement, and understanding. Without him, this enterprise would have had little meaning.

THE BARNS OF THE NORTH FORK

Southold

The original barn was built in the mid-1800s.
Adding a second barn to the first with the utility of a shared wall occurs throughout the North Fork. As in this instance, the second barn is often smaller than the original.

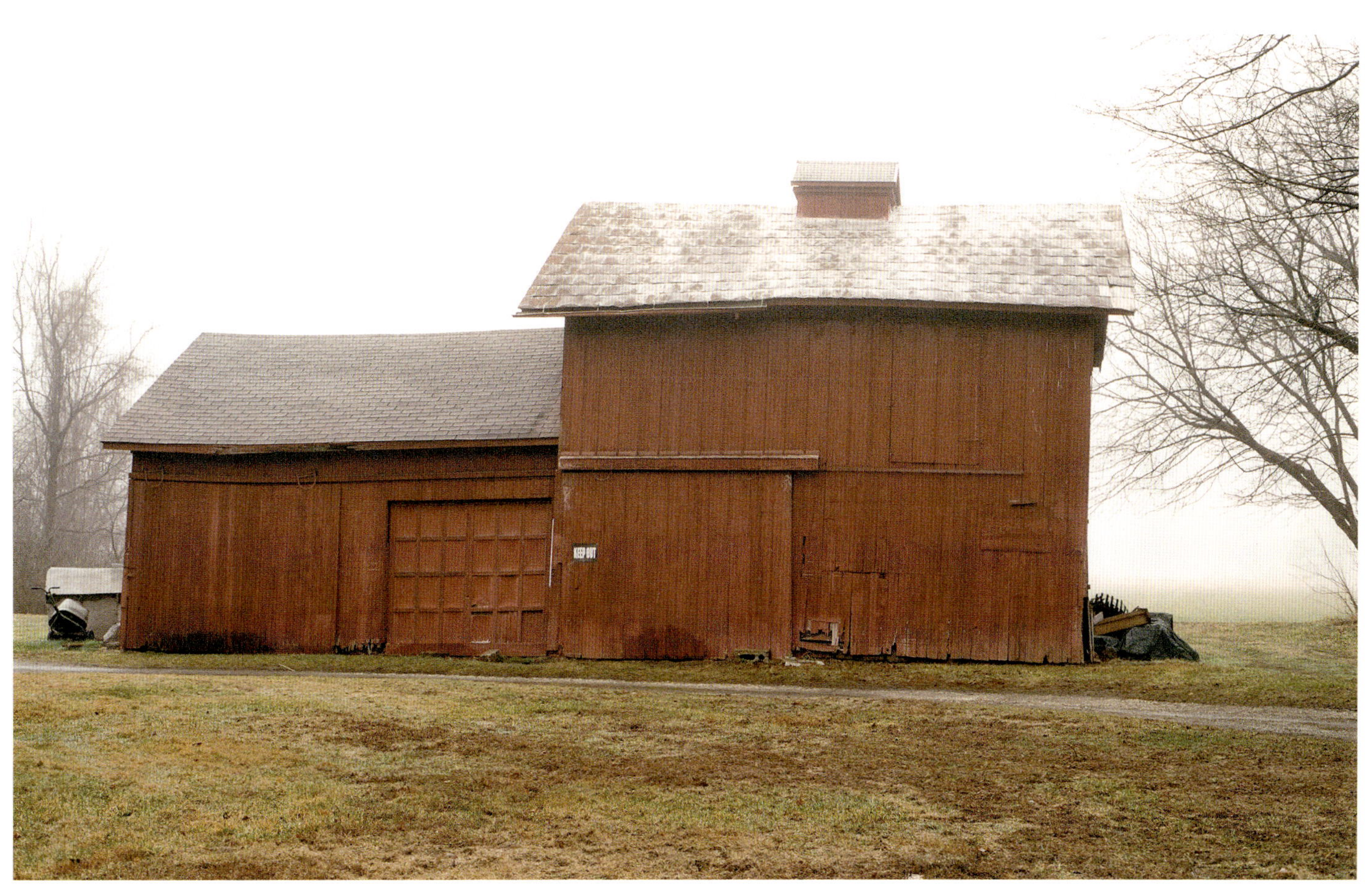
KEEP OUT

Peconic

Barns are visual icons in the rural landscape.

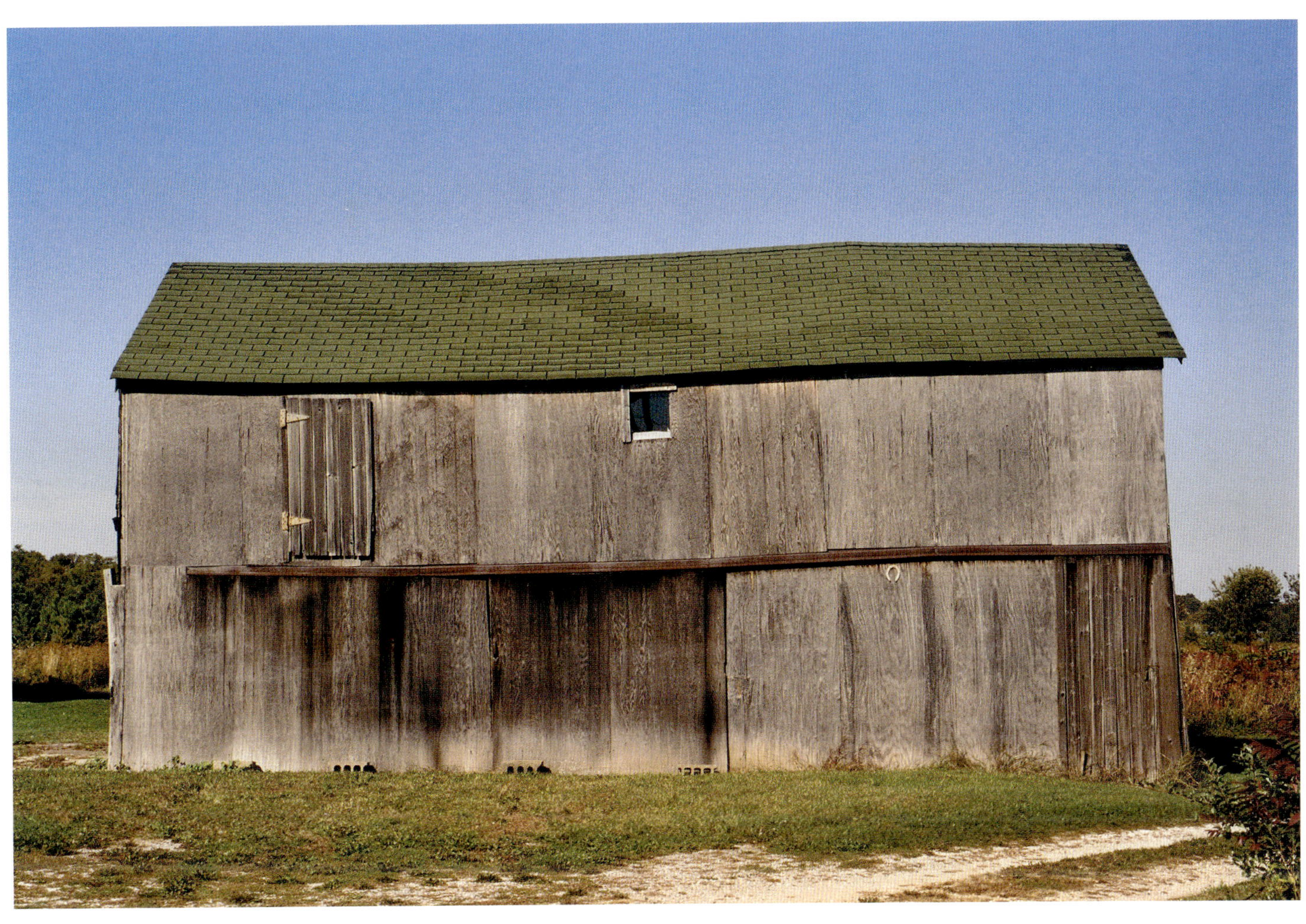

Peconic

Two bays were added to the original three-bay barn for a total of five bays. The older barn's frame is hand hewn, mortise and tenon; the rafters are pinned and there is no ridge pole. The added bays have a ridge pole and the rafters are not hand hewn.

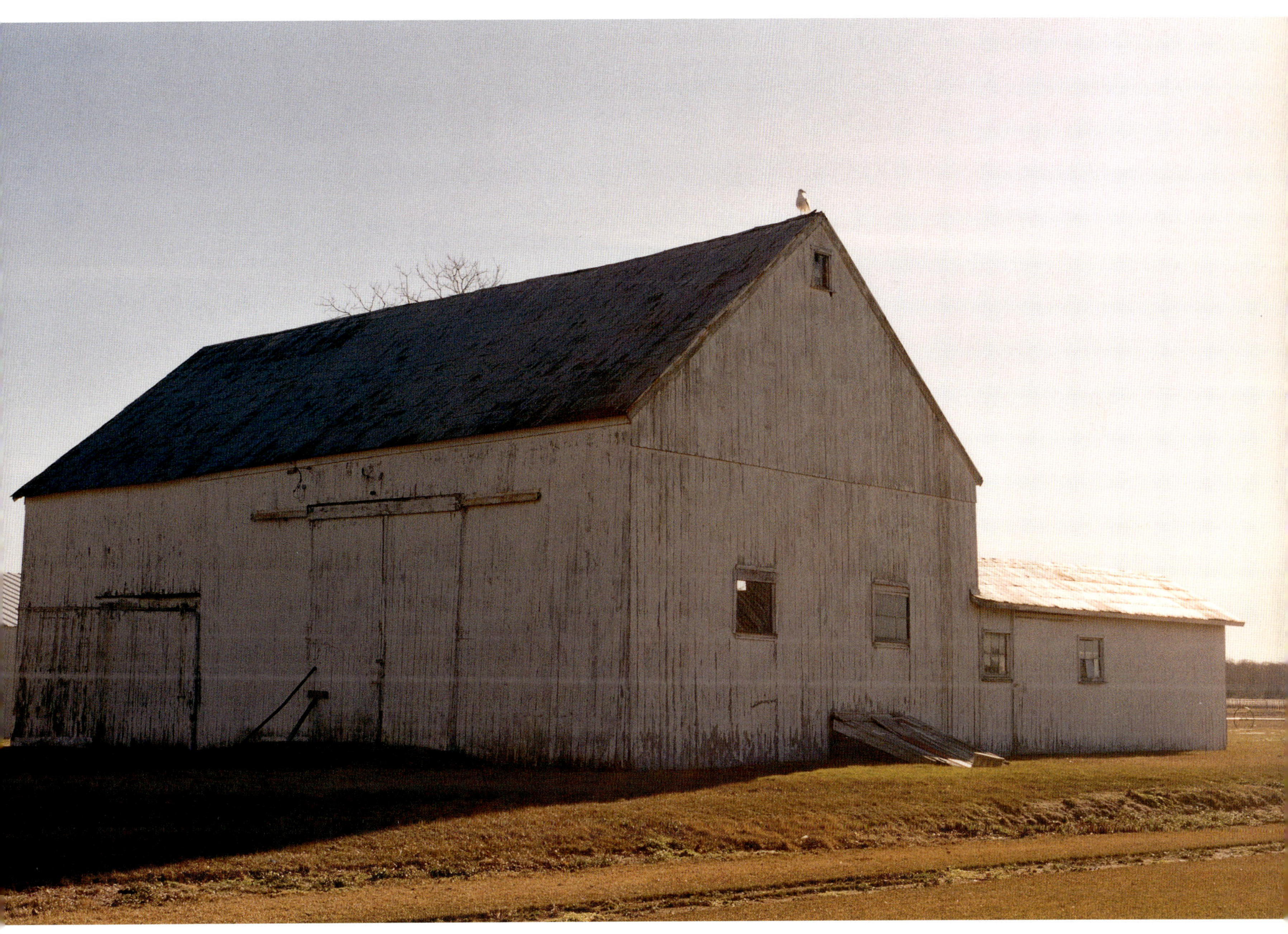

Cutchogue

This form—main barn with shed addition—is quintessential Long Island.

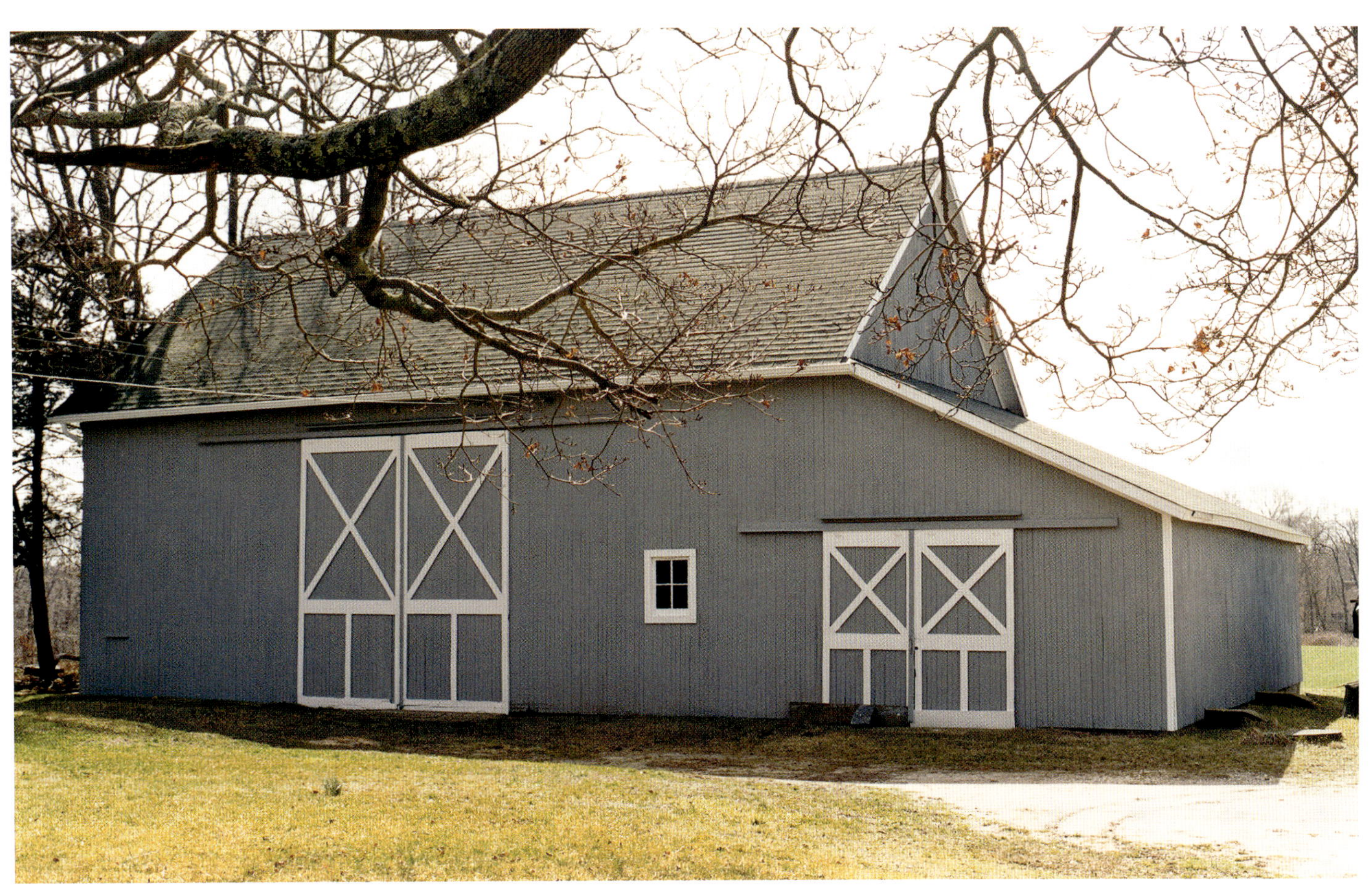

Riverhead

Seen in the early morning or across
the field a barn has a spirit of its own.

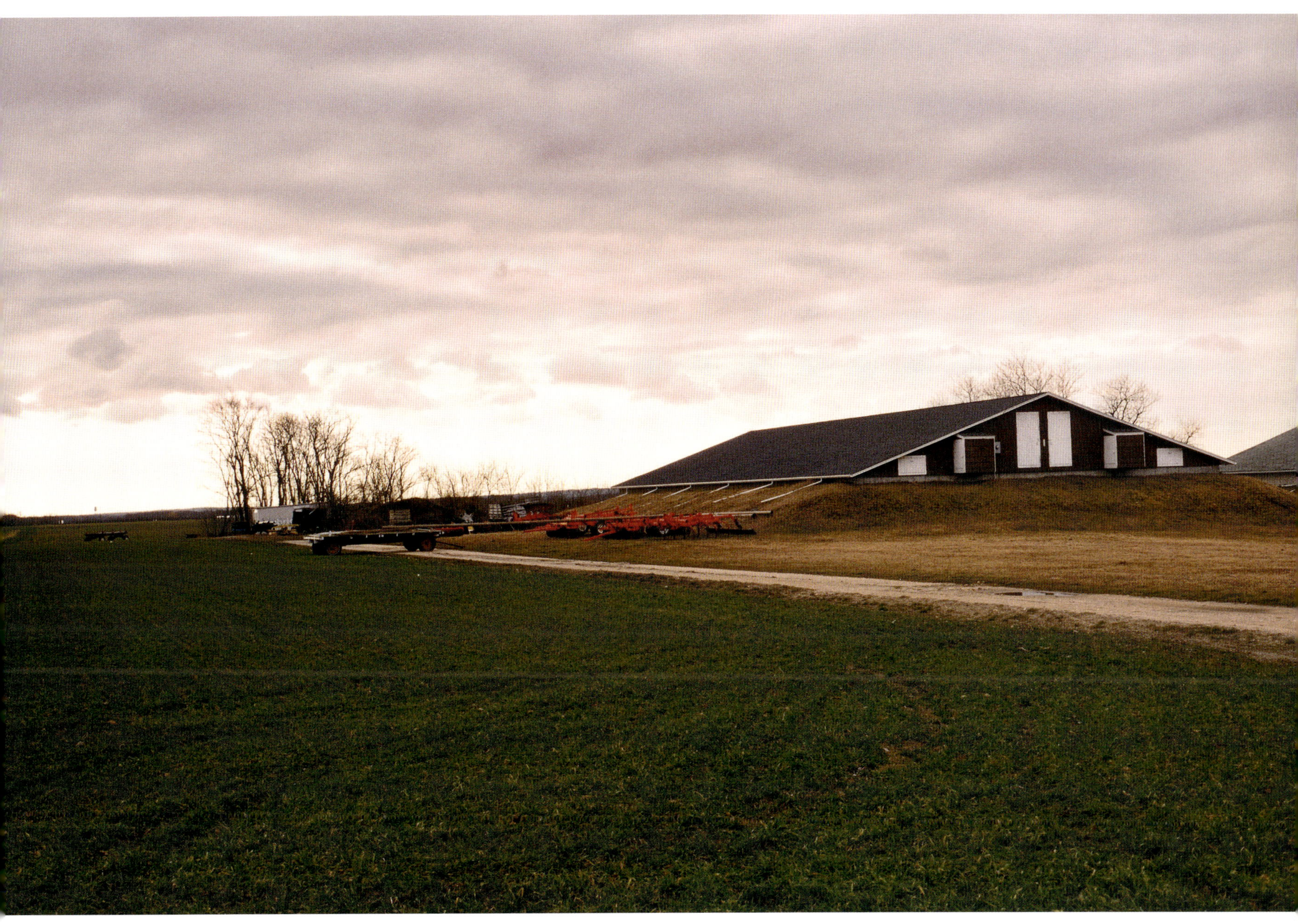

Oregon

The two-story double barn was built in 1890 and has been renovated for use as a tasting room. The asphalt-shingled barn was built in 1930–40 and is in agricultural use.

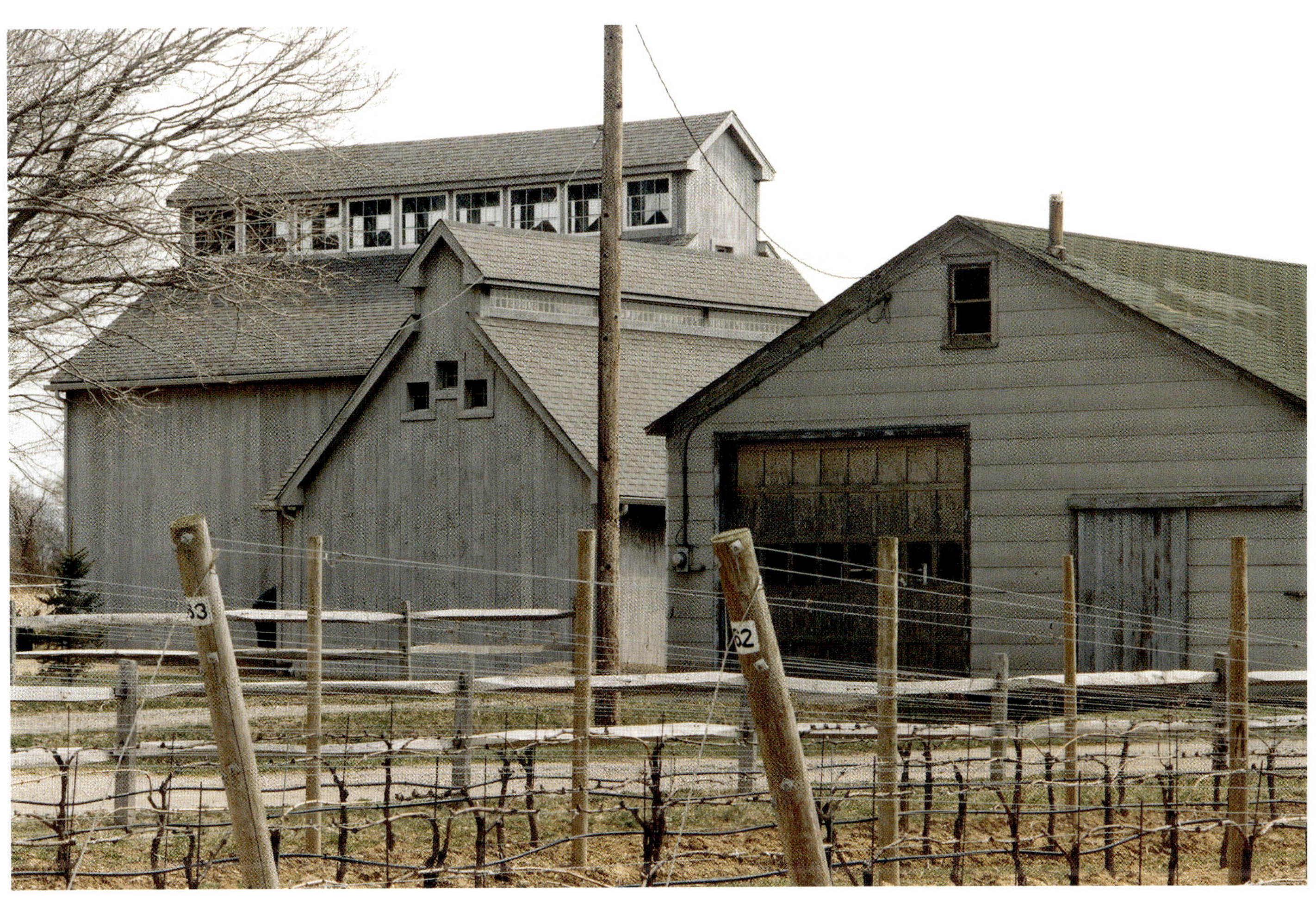

Bayview

Built in 1908 by Haven Emerson as a dairy barn, this barn's terra-cotta brick exterior is unique. The three-story barn is also unusual because it is built into the bank of a hill. Bank barns—as opposed to potato barns—are rare on the North Fork.

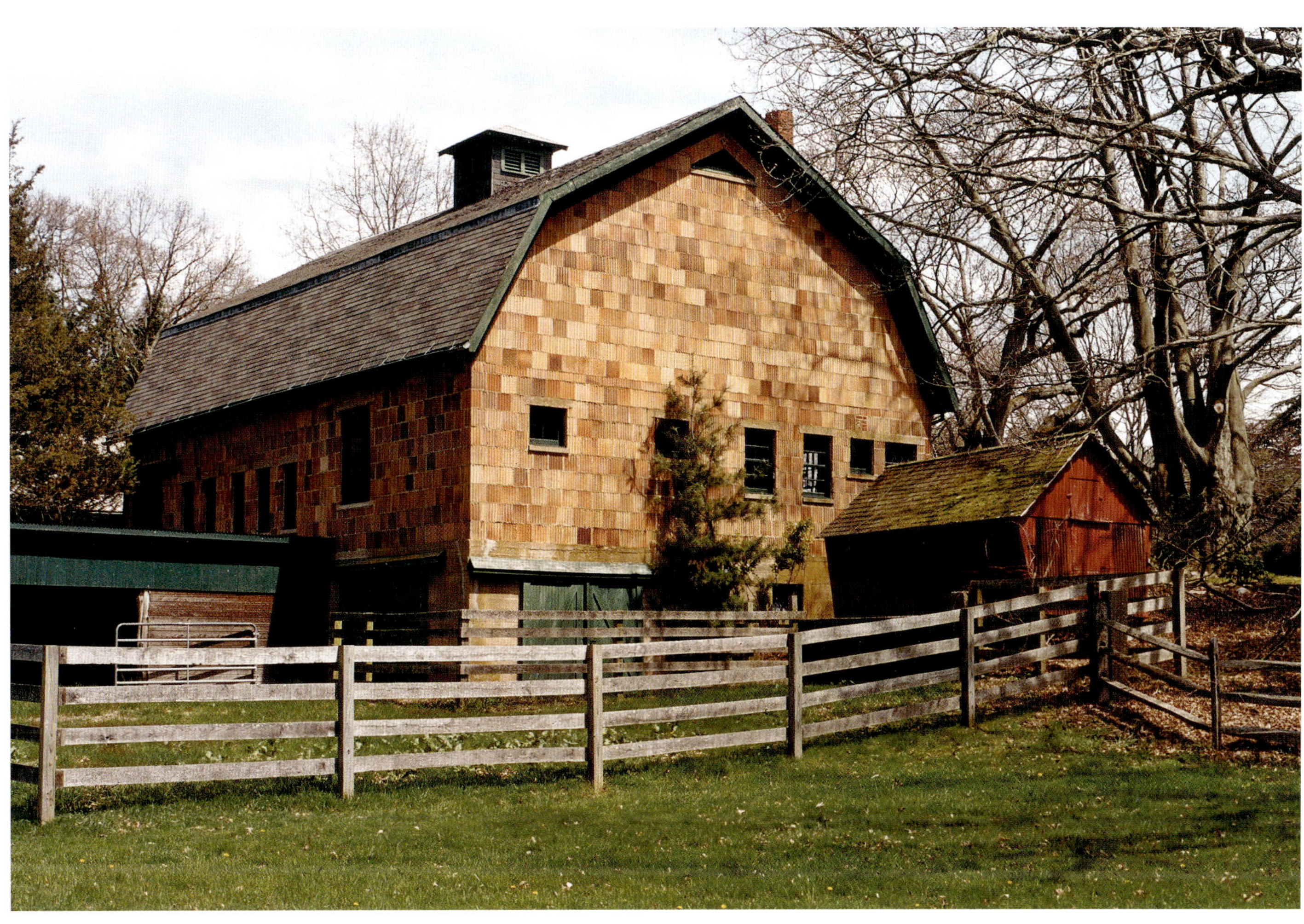

Greenport

Small barns appear in residential settings. They may have survived from an agricultural past or they may be carriage houses.

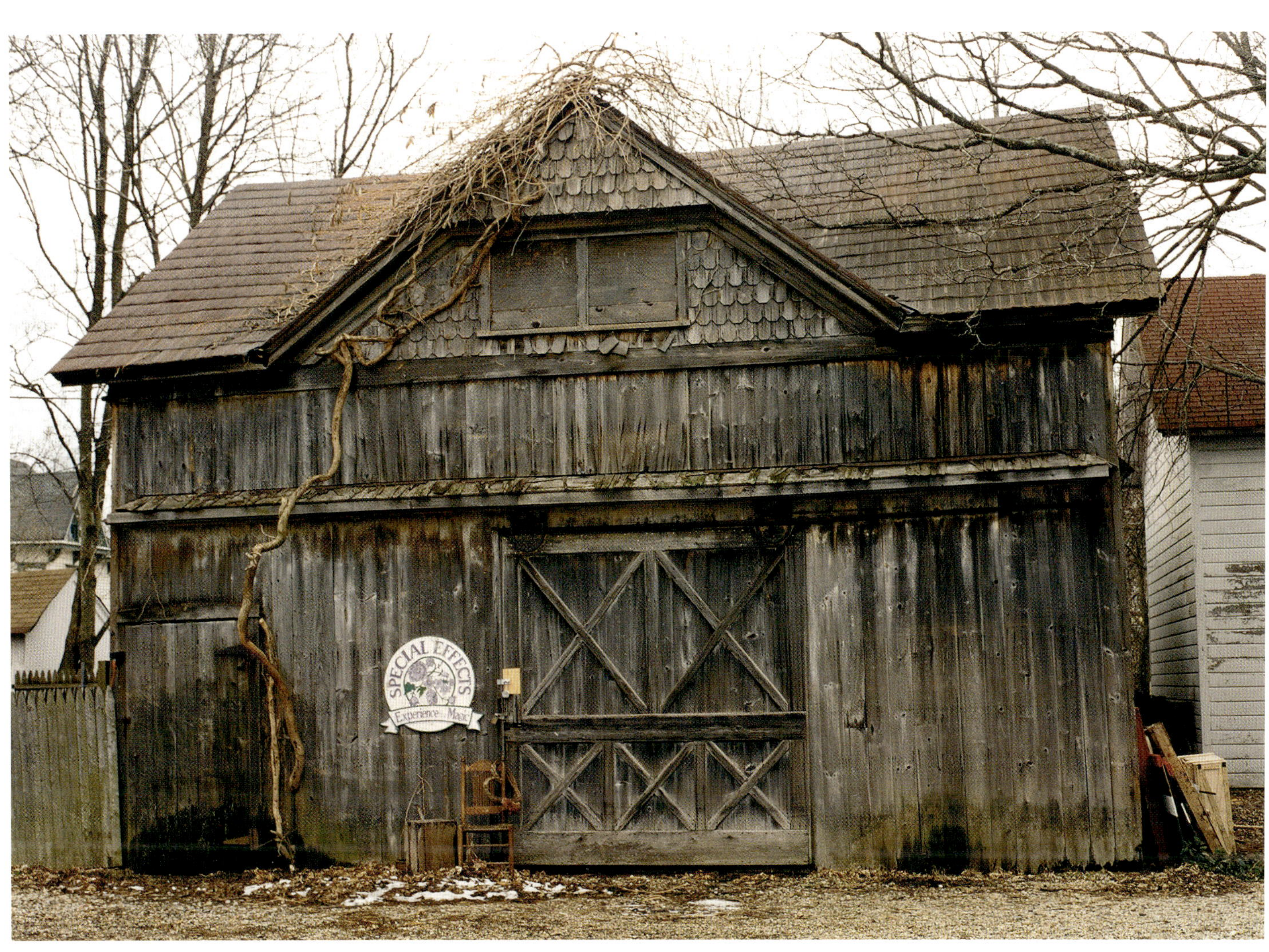
SPECIAL EFFECTS
Experience Magic

Mattituck

The long, five-bay barn was built in the nineteenth century.

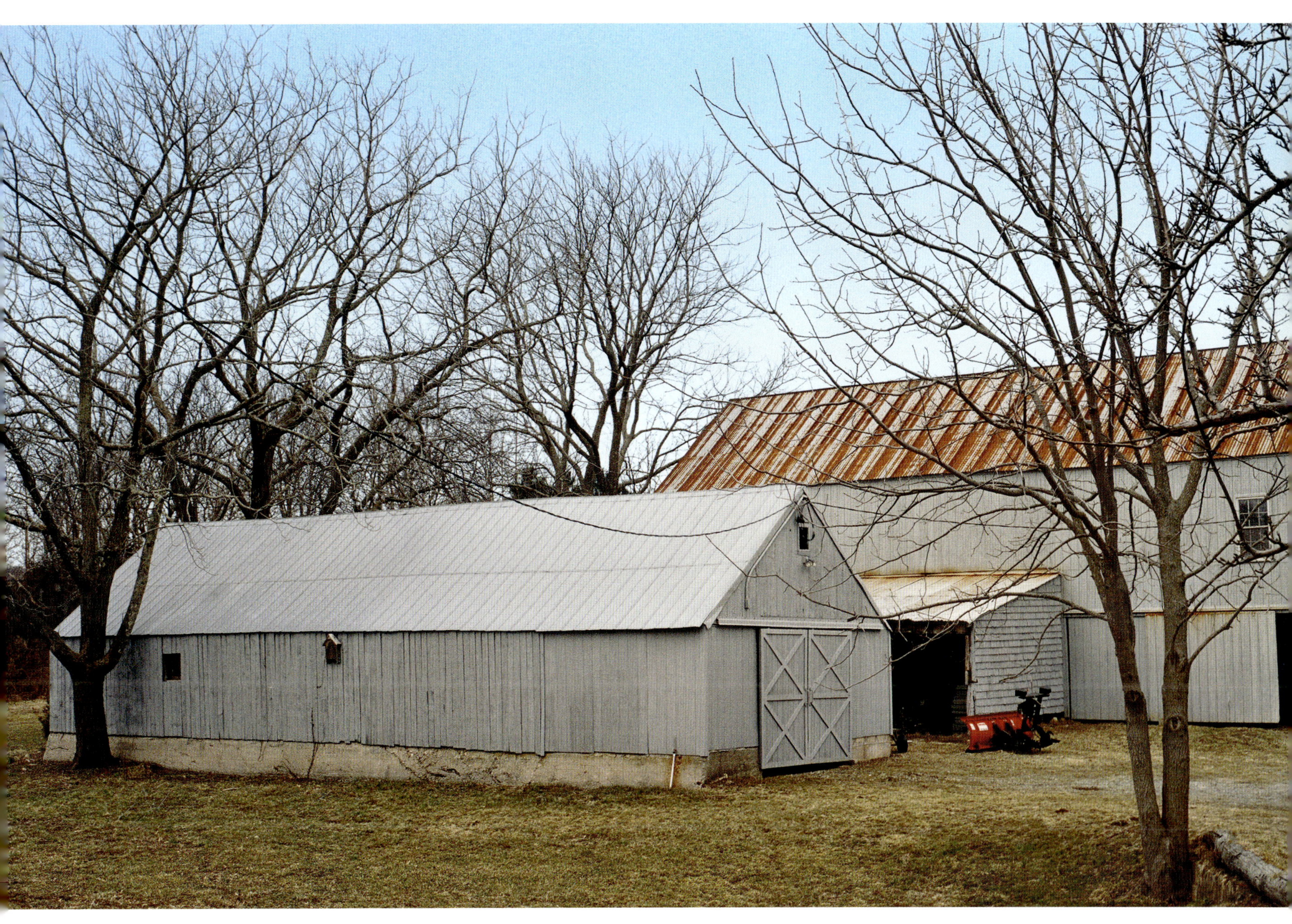

Bayview

The barn has a basic English hay barn configuration and was extended on the southern end by 6–8 feet. It was built in the nineteenth century.

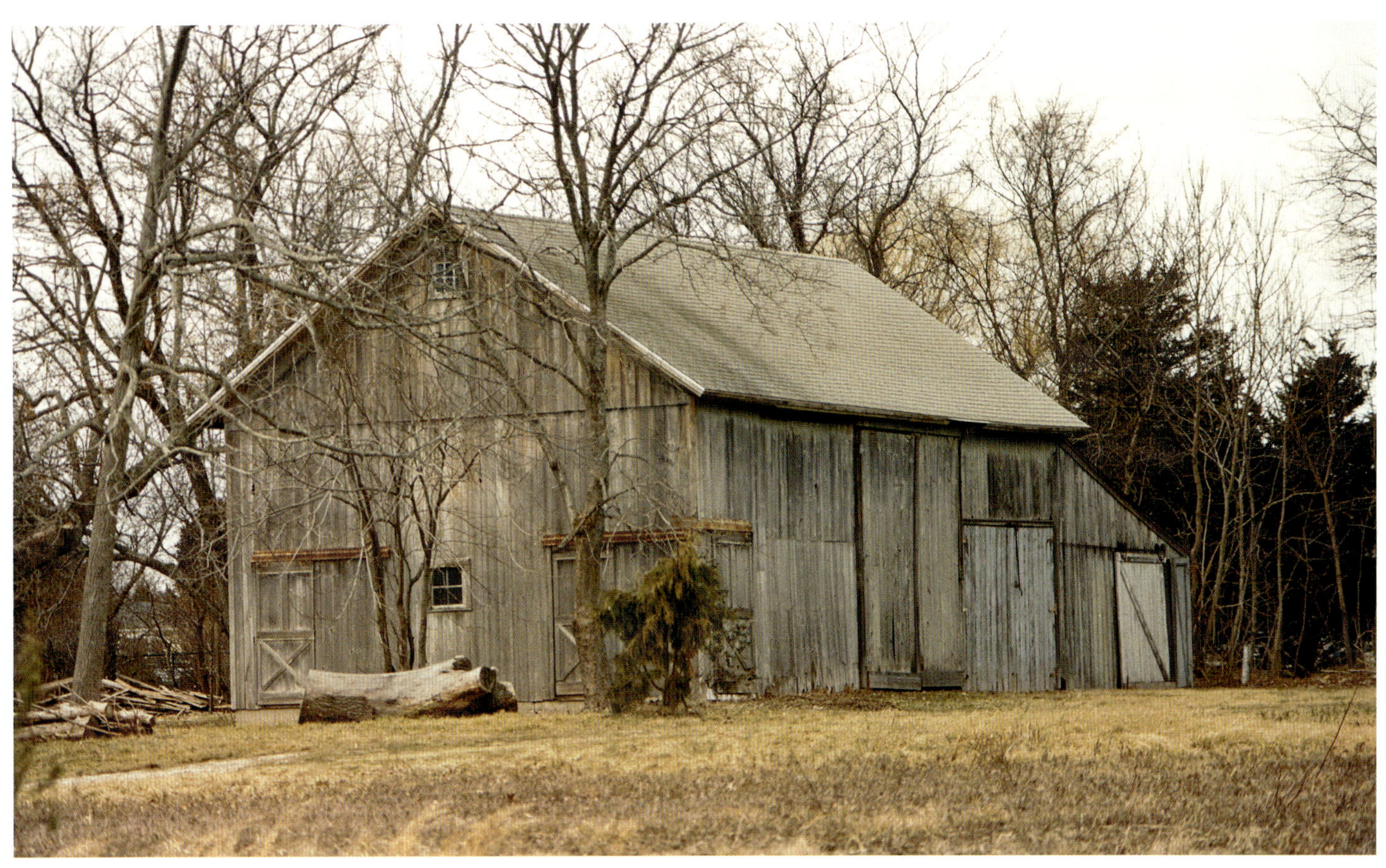

Cutchogue

Barns illuminate the evolution of farming over time.
This barn complex is now part of a vineyard operation.

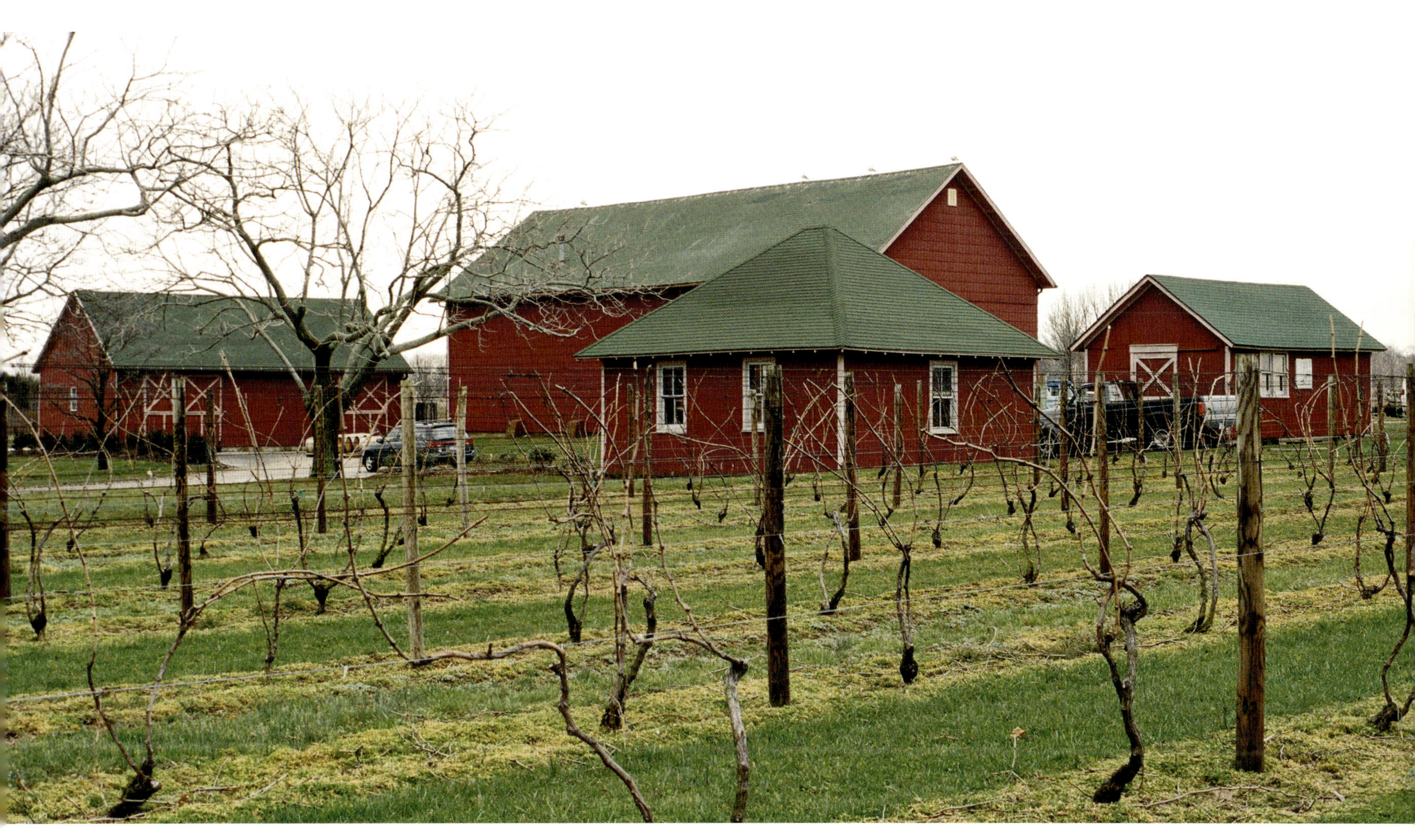

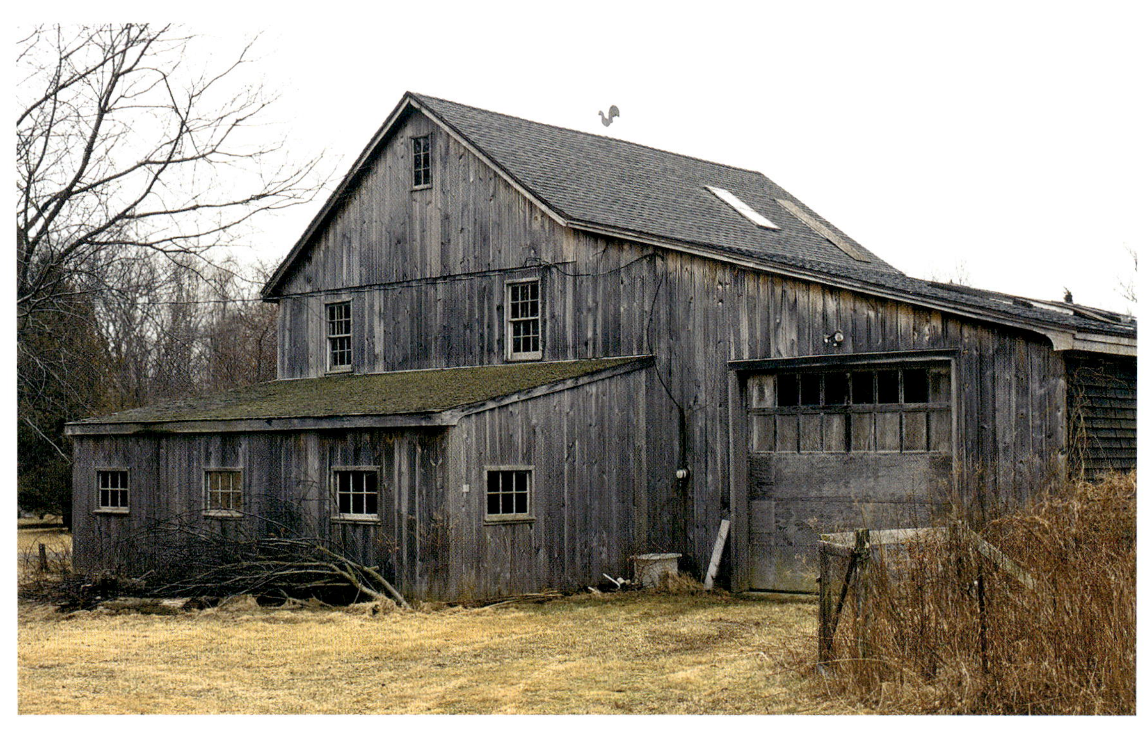

Mattituck

Barns need to be functional, which accounts for their expansion by means of wings or shed additions.

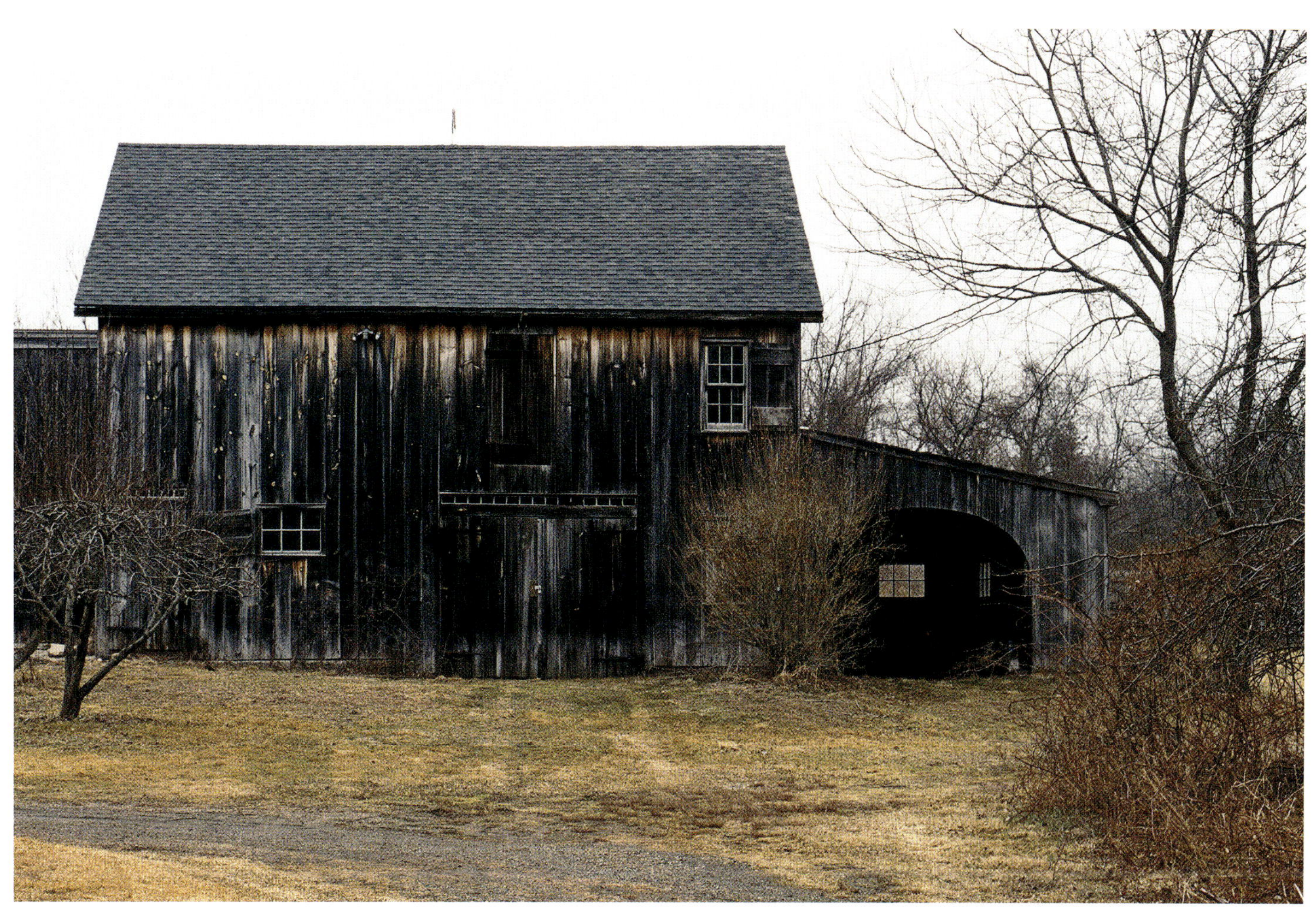

Orient Point

The main barn was moved to its present site from the adjacent property sometime before the 1820s. It is an English three-bay barn and the major portion of the framing appears to be from the eighteenth century. A wide board floor in the center bay has been preserved.

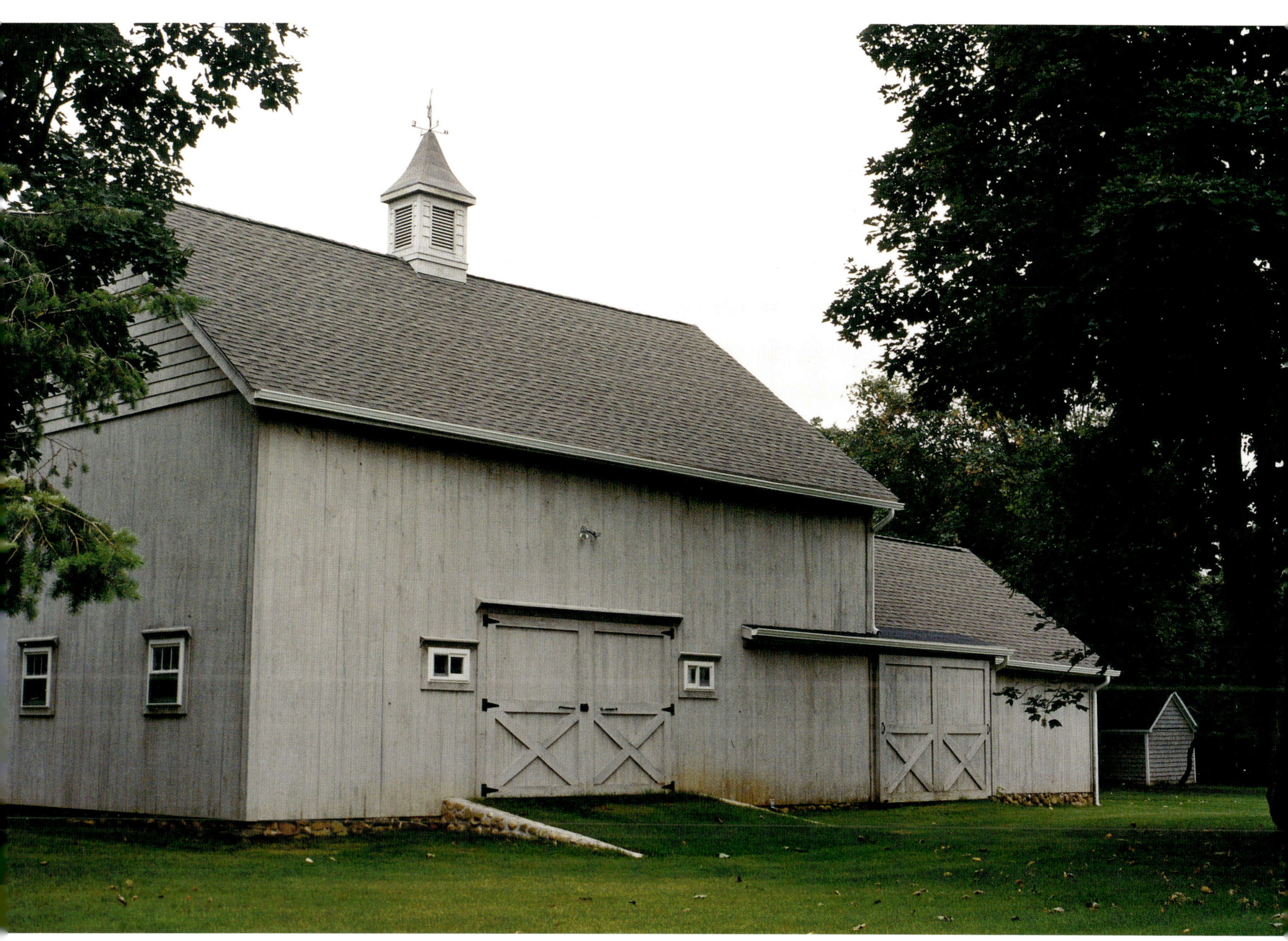

Greenport

These are former dairy barns.
Silos are few and far between on the North Fork.

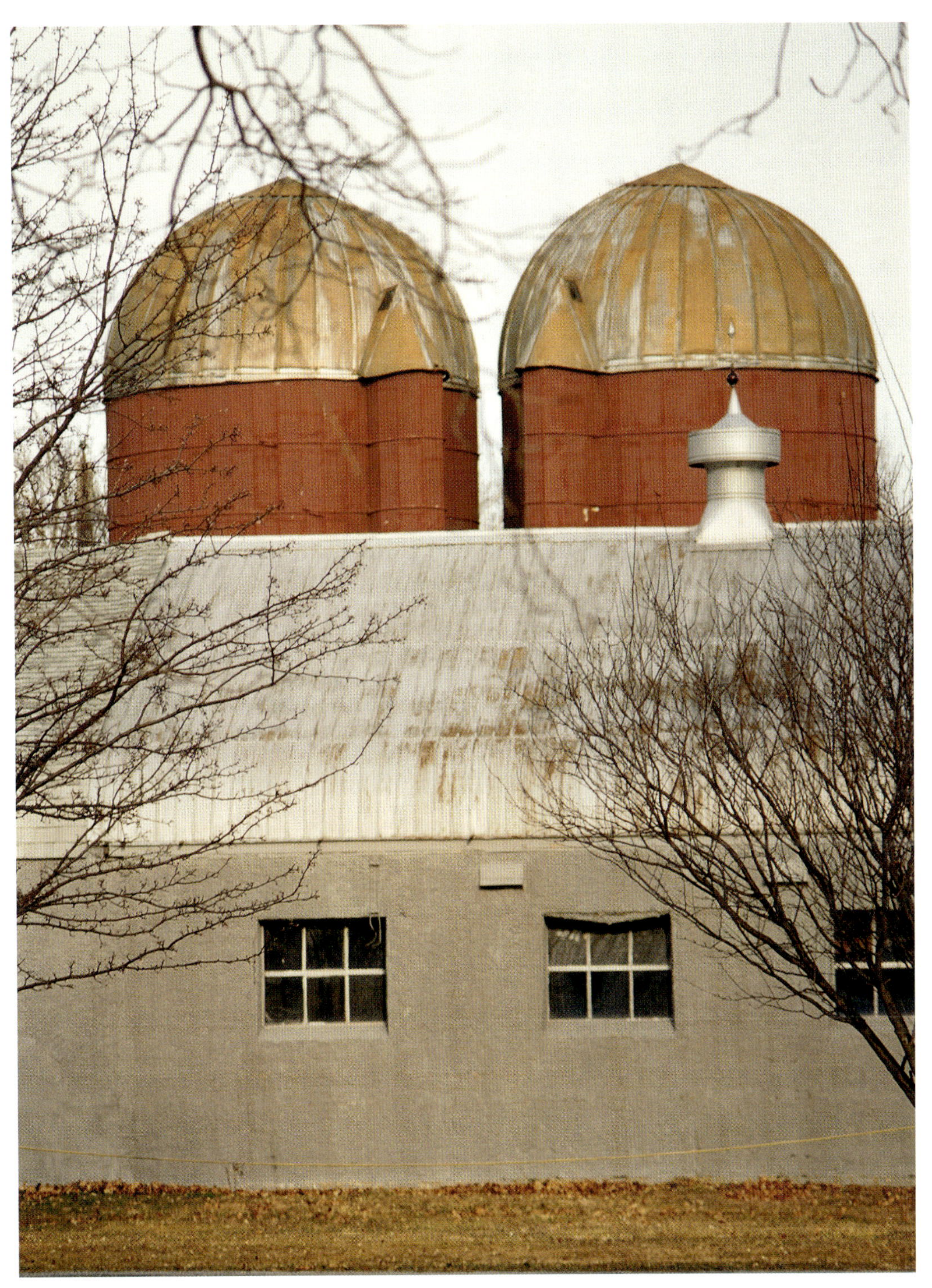

Mattituck

The roofline and fenestration of this barn suggest that it may have originally been a three-bay barn to which a later bay was added.

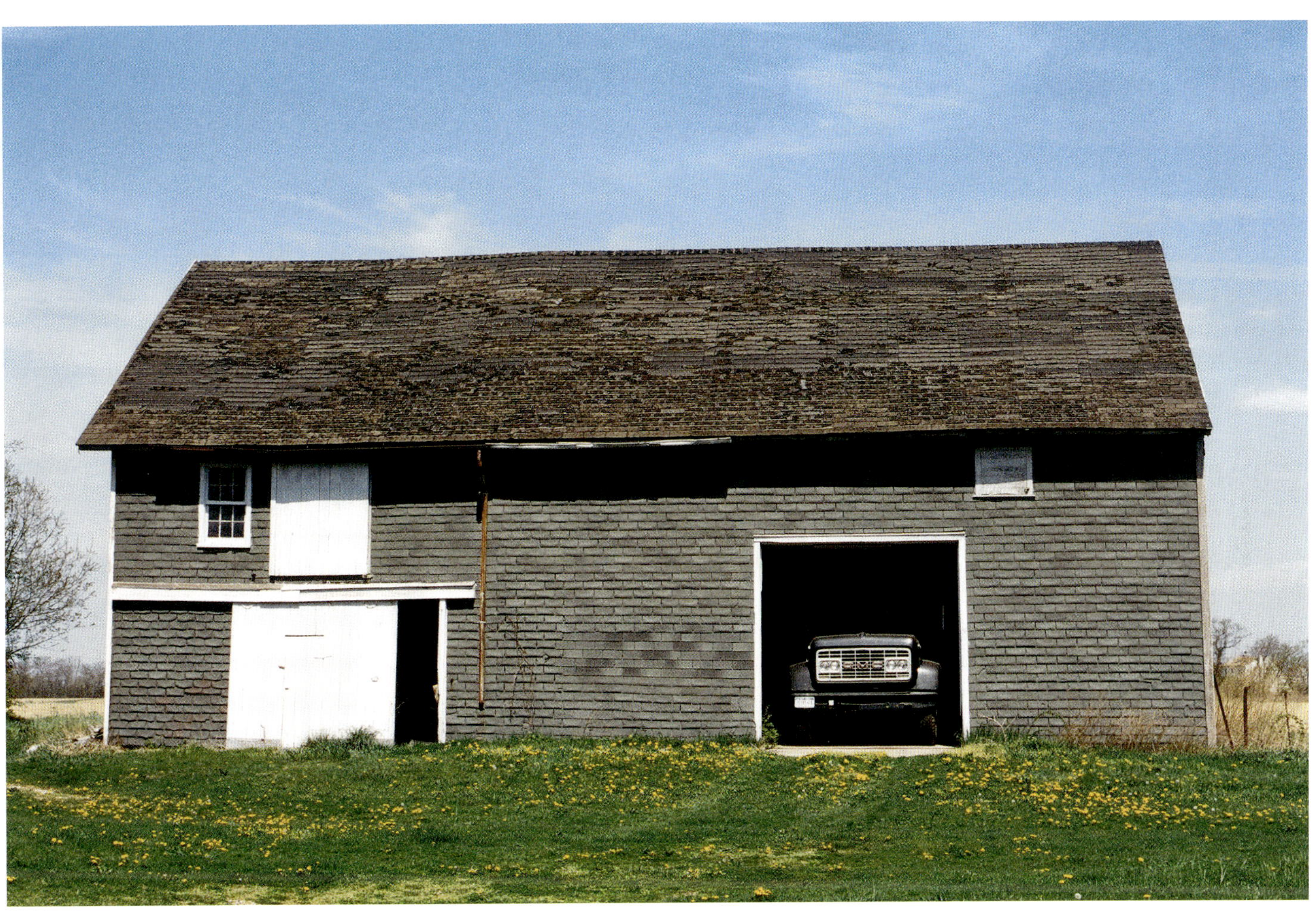

Oregon

Shed additions follow no formula. They are attached singly or flanking as on this barn. Sometimes they surround two sides like a wraparound Victorian porch.

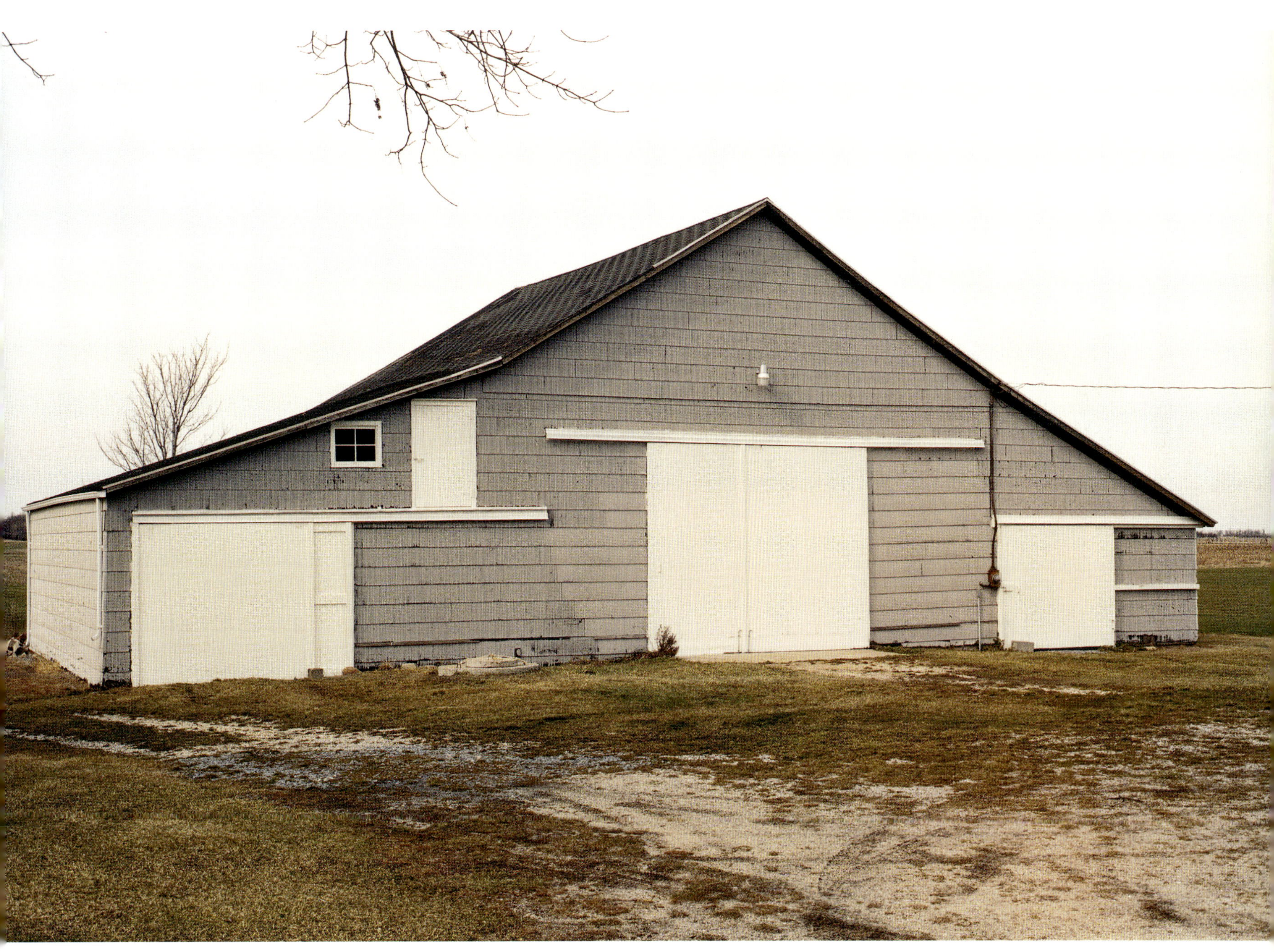

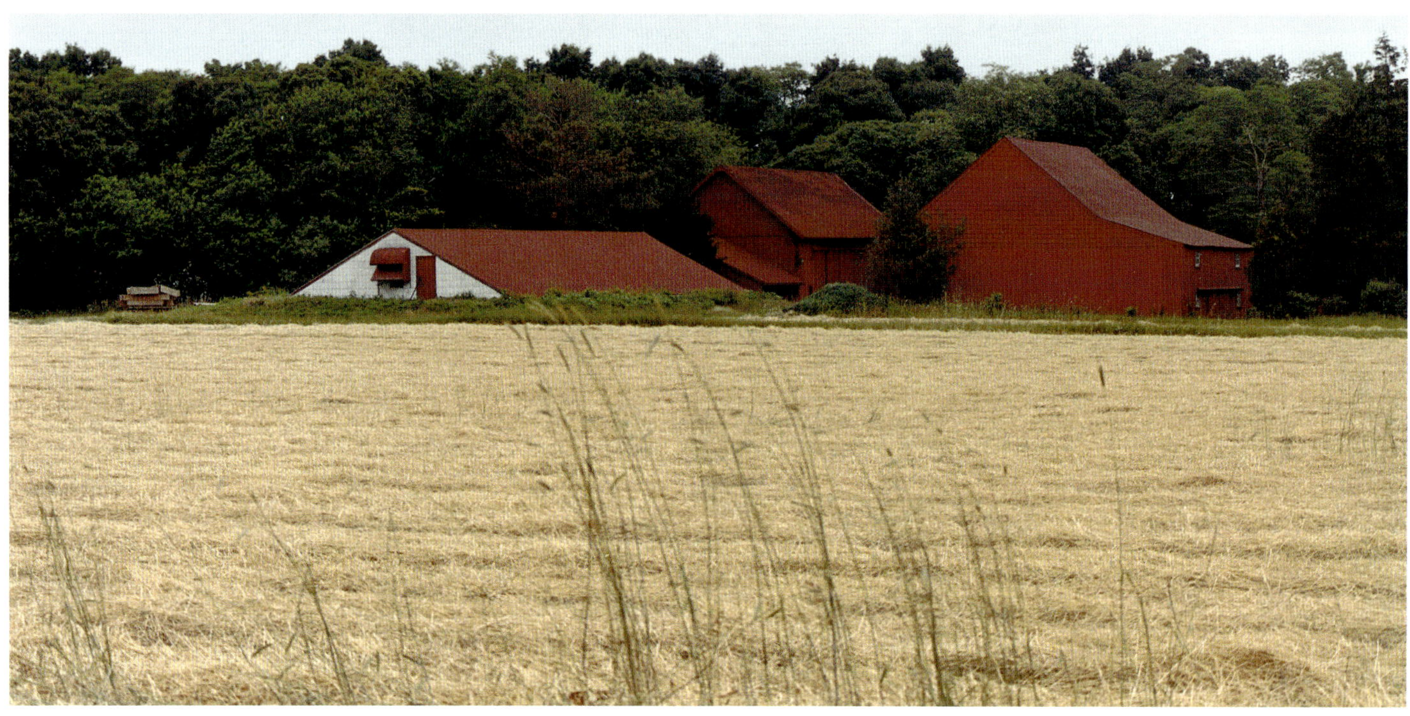

Cutchogue

This set of barns represents three centuries of barn building. The English hay barn was built in the eighteenth century, the double barn is from the nineteenth century, and the third barn is a potato barn built in the twentieth.

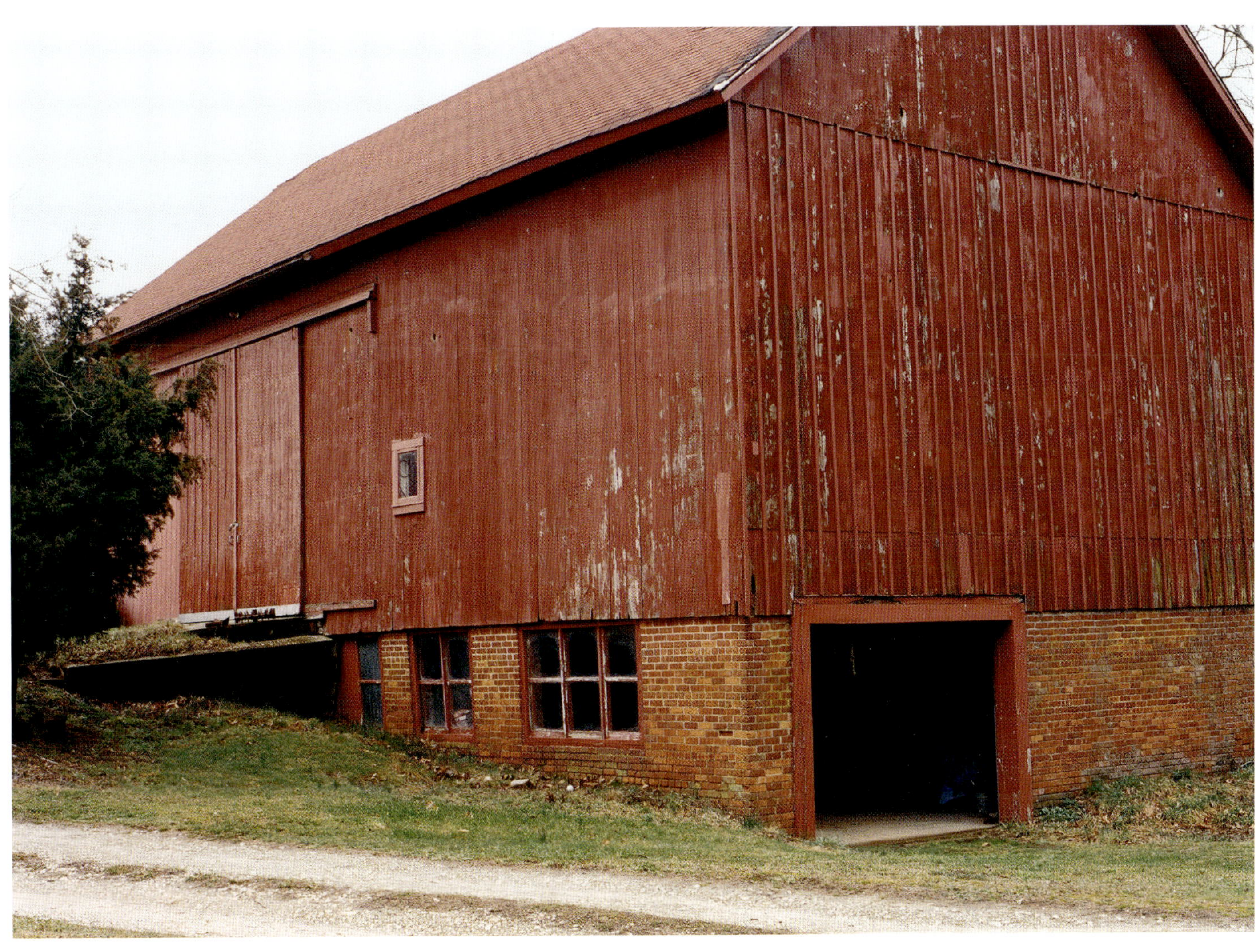

East Marion

This three-gabled barn is more than one hundred years old. It belonged to Captain Leak. The shadow of a gable suggests that there was an additional attached structure at one time.

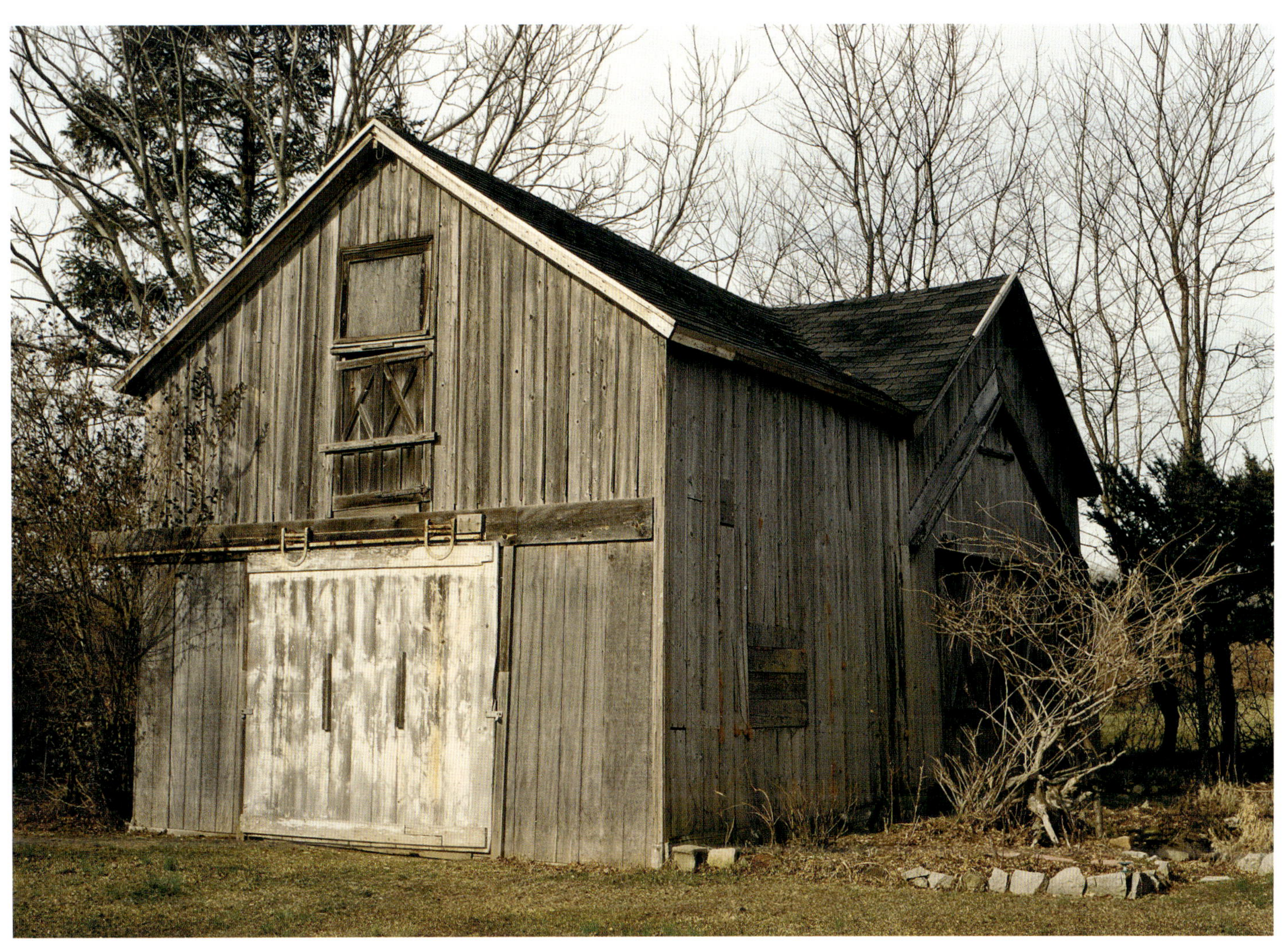

Orient Point

Both the square form and the door on the gable end of this barn are unusual.

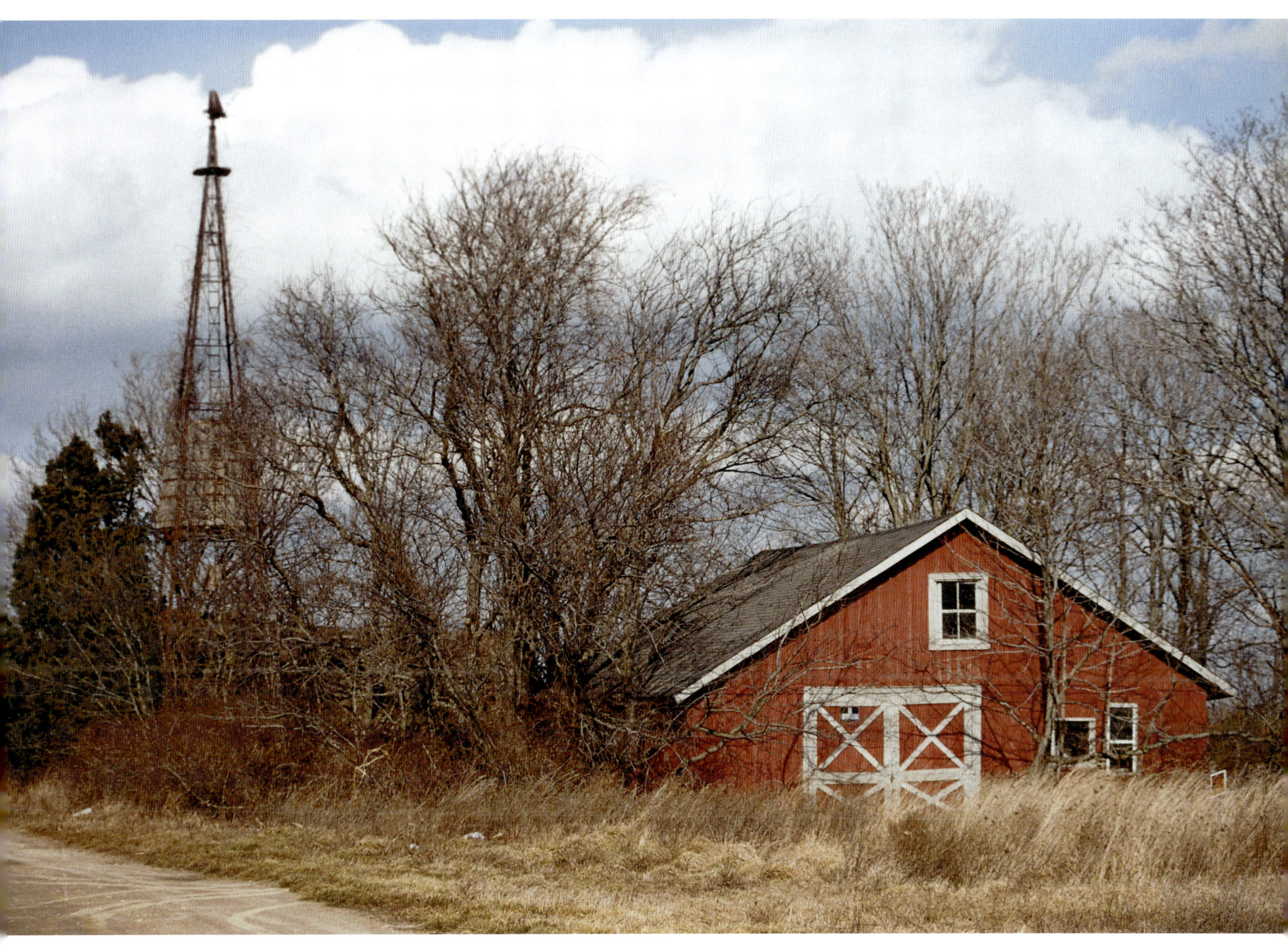

Cutchogue

The barn was built in the 1920s along the railroad tracks. It is a potato barn constructed with steel beams. Potatoes were conveyed to the second floor. Three railroad cars at a time could be loaded from the barn.

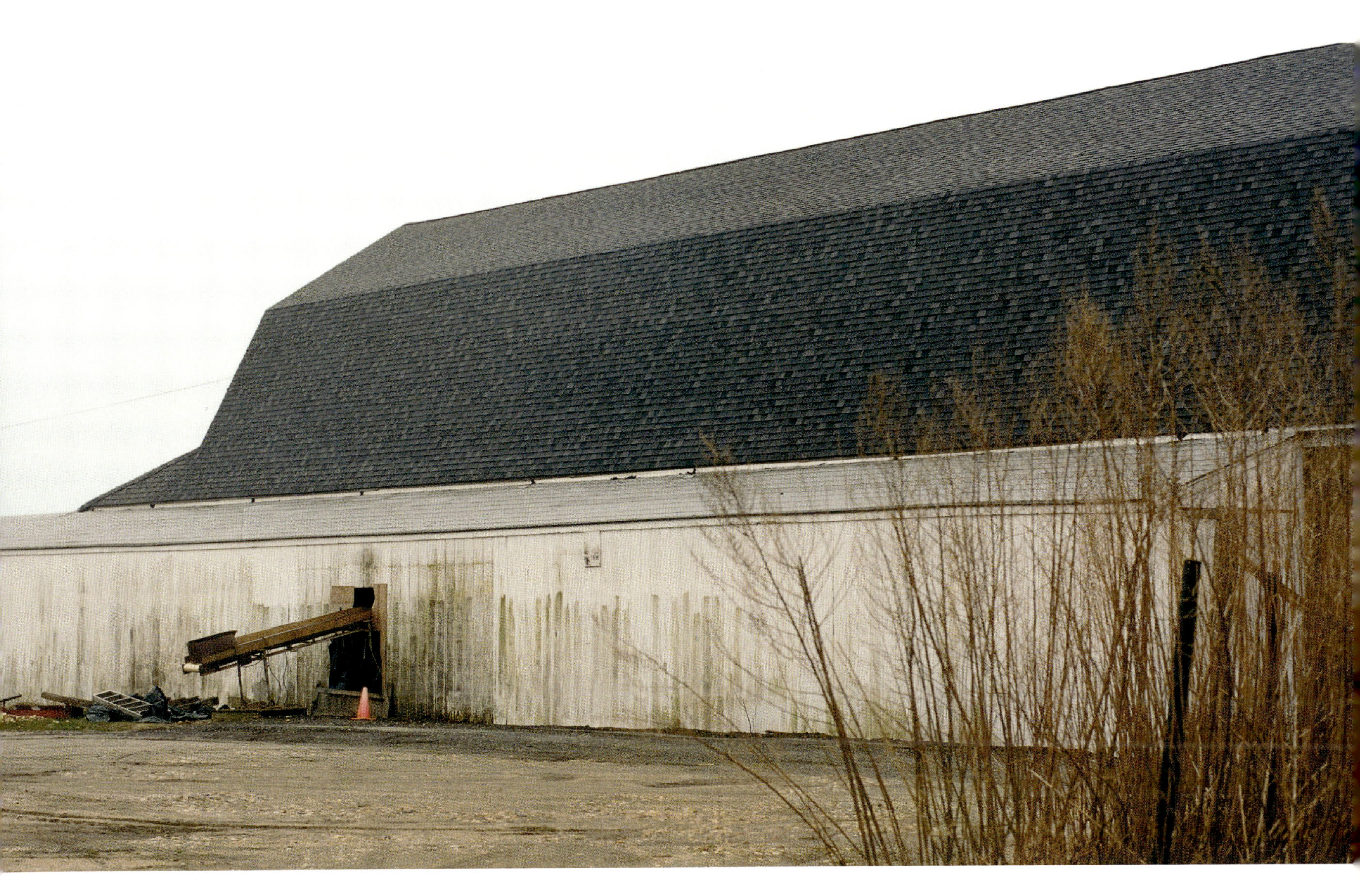

Oregon

The barn was built circa 1890.

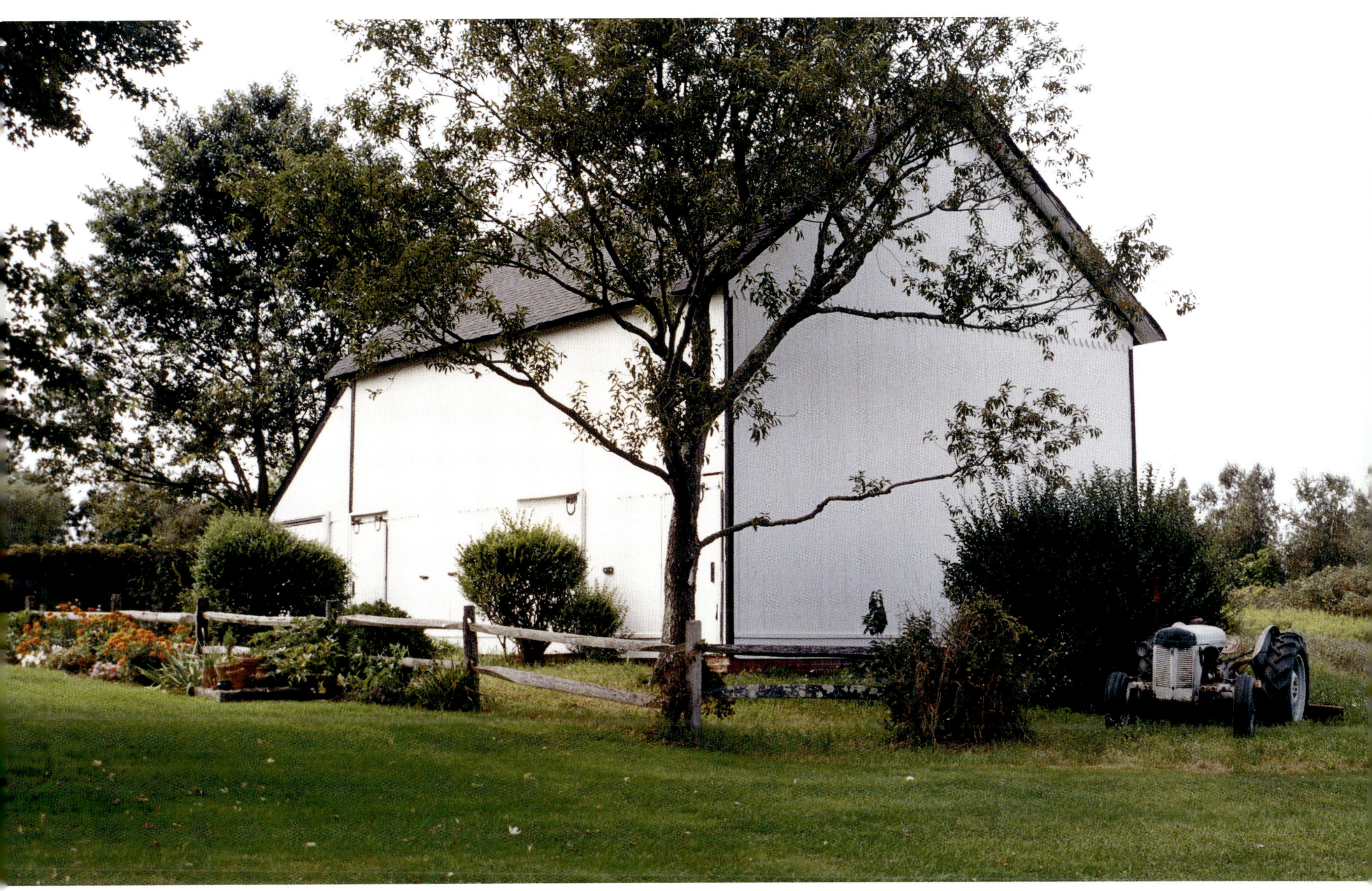

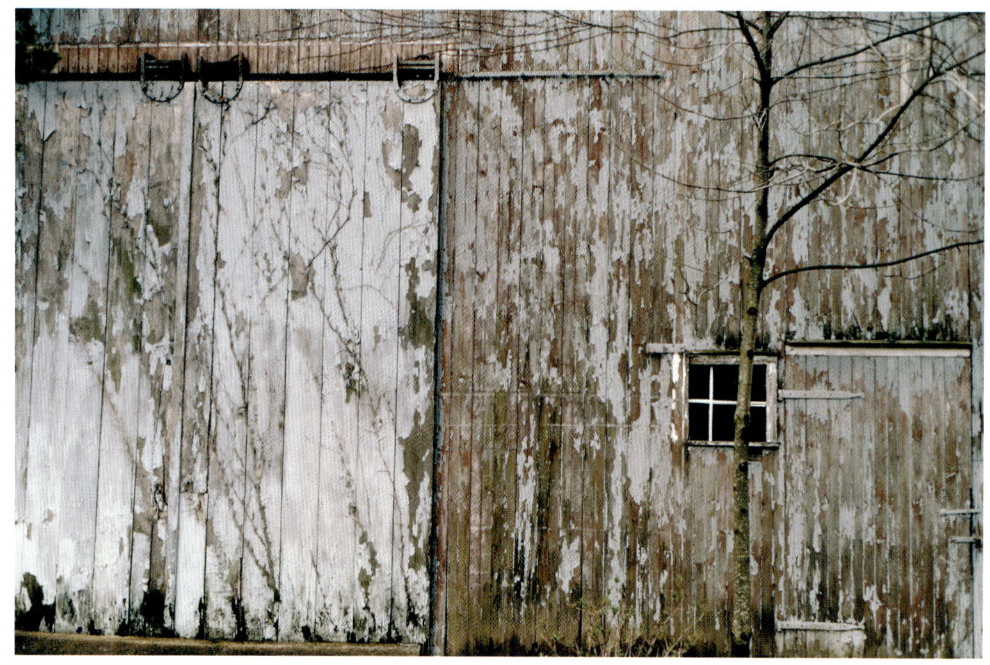

Southold

This English three-bay barn with hand-hewn timbers and mortise and tenon joints was built in the nineteenth century or possibly earlier.

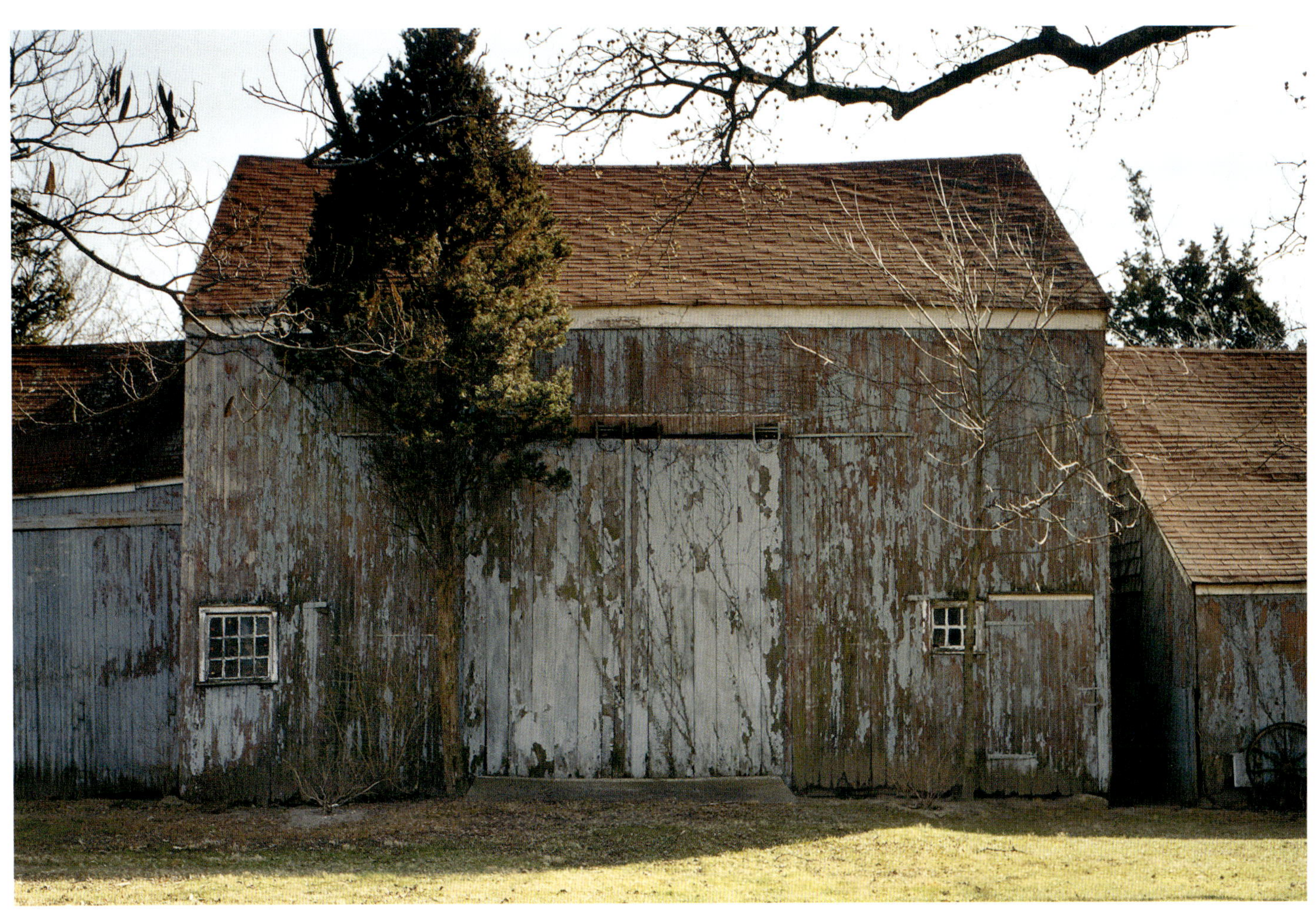

Riverhead

Are these outbuildings or barns? Usually small farm buildings are labeled by their function, such as cow shed.

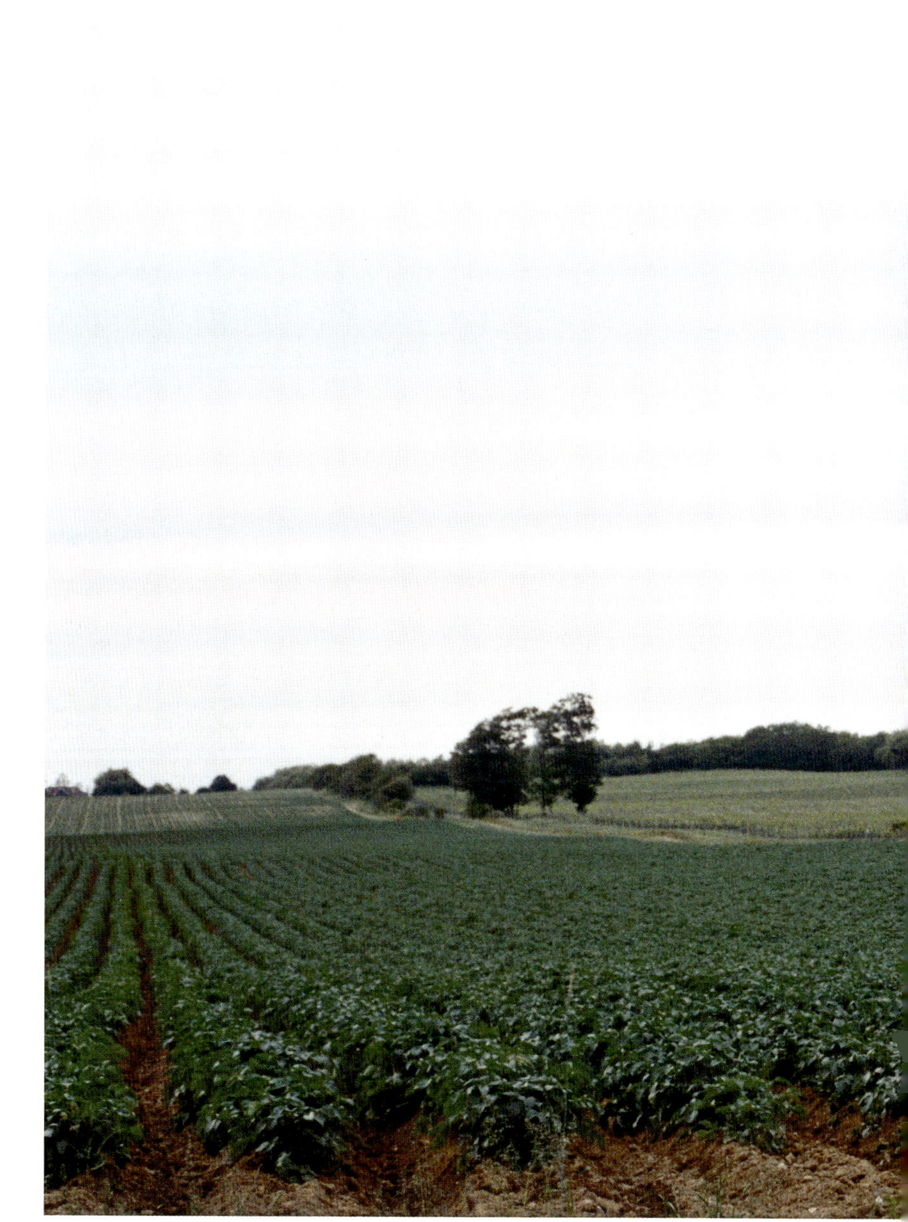

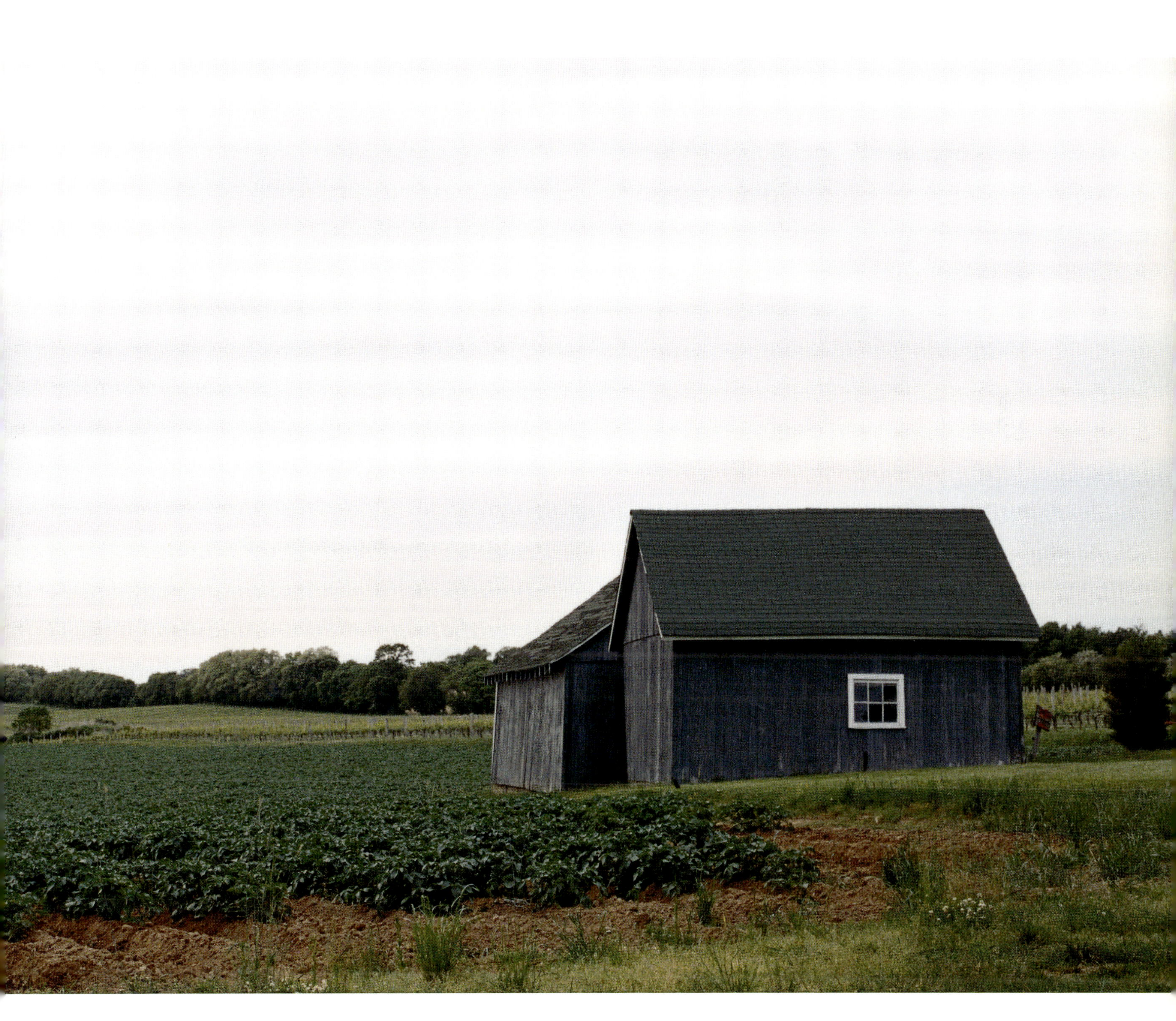

Orient

Portions of the frame of this barn are pegged and there are framing marks. These elements may come from an earlier barn.

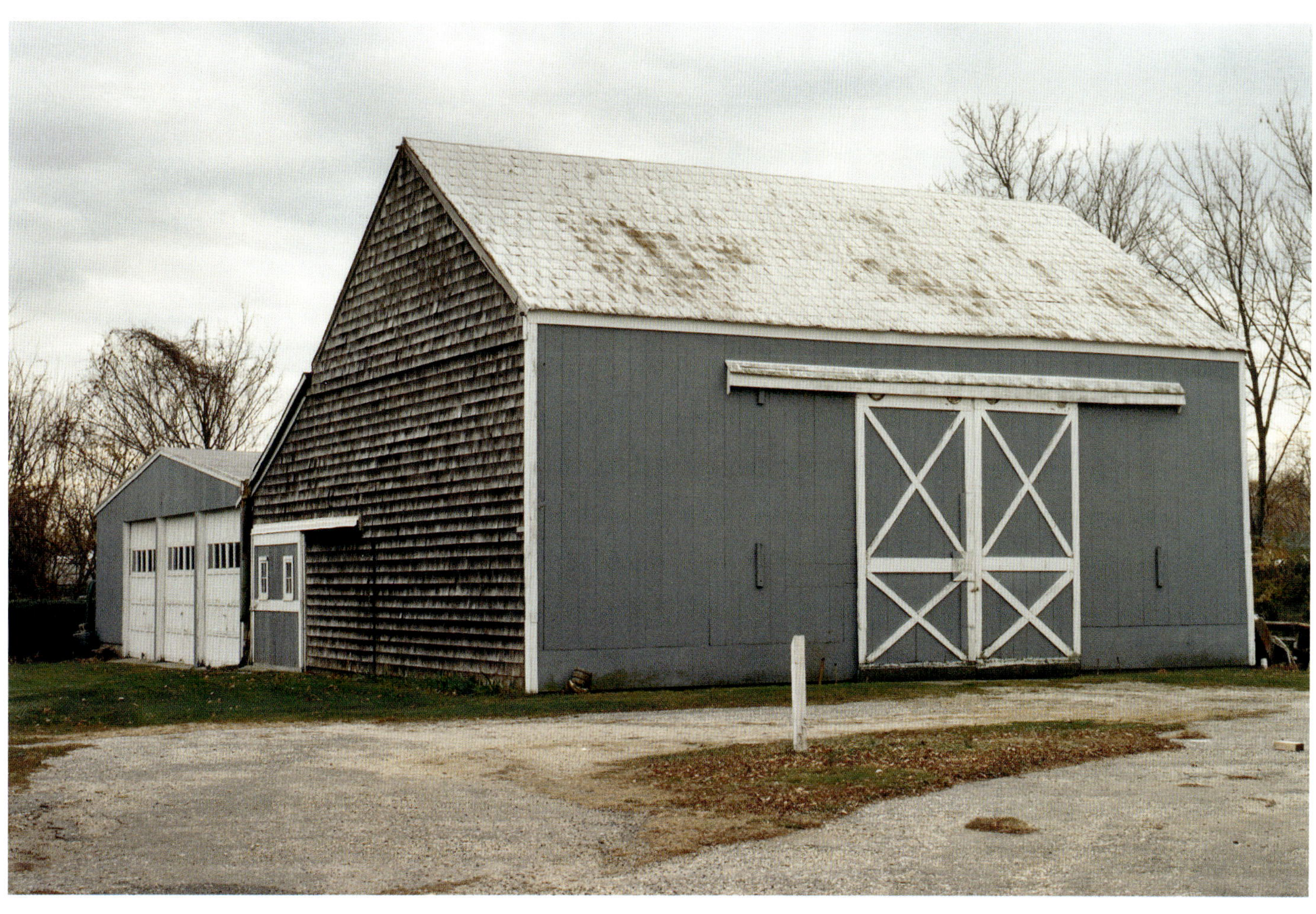

East Marion

East Marion was a fishing village and baymen's shacks line the Main Road. Fishermen used them for tarring and mending their fishing nets. They are the size and shape of carriage houses. The small shed on this structure contained an iron well, which was used for tarring nets.

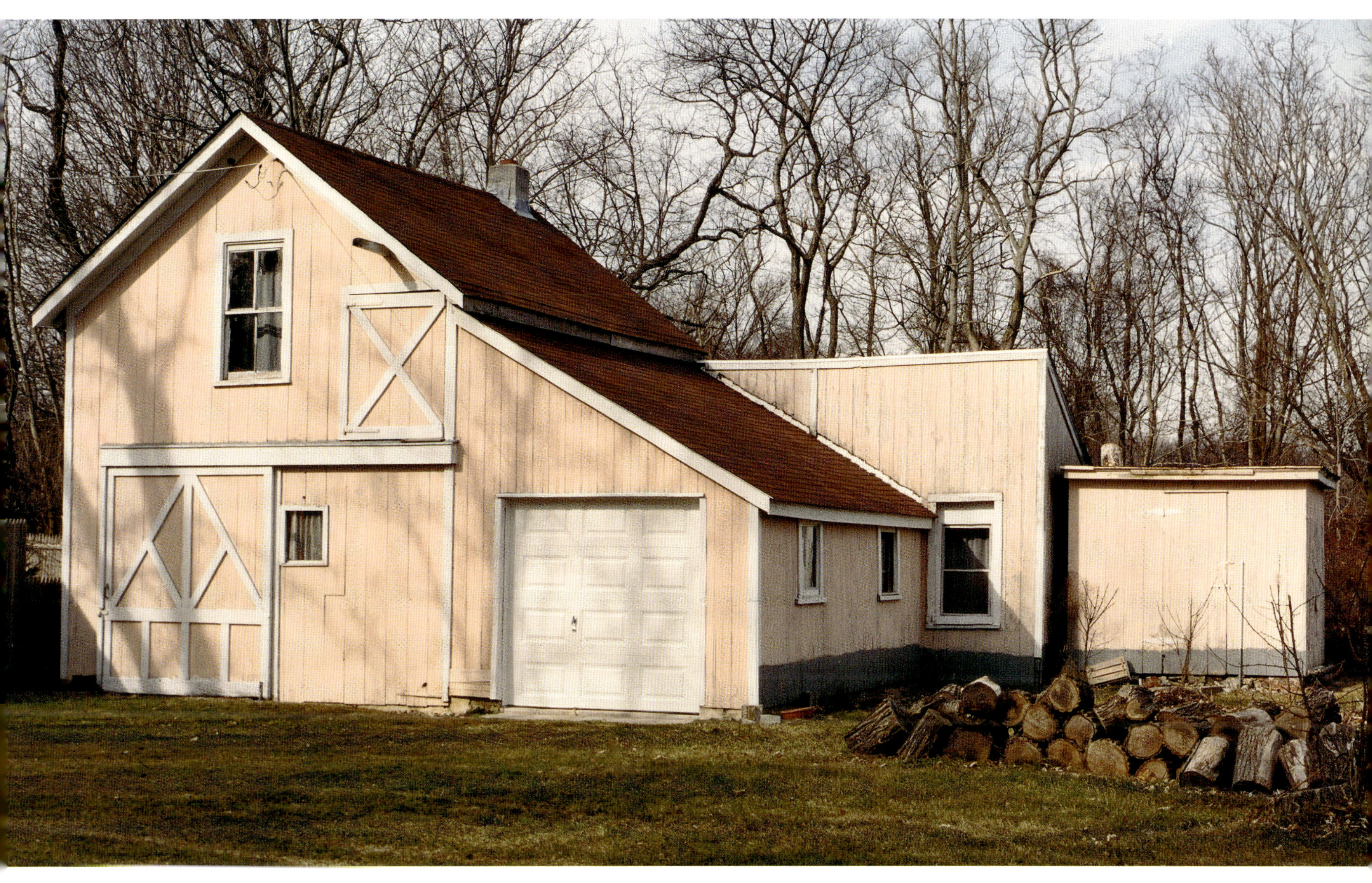

Mattituck

The upper portion of this barn was Mattituck's second schoolhouse, built in 1857. It has been raised one story (brick) and flanked by wood-shingled shed additions. The first Mattituck schoolhouse (built before 1829) sits alongside the barn to the east.

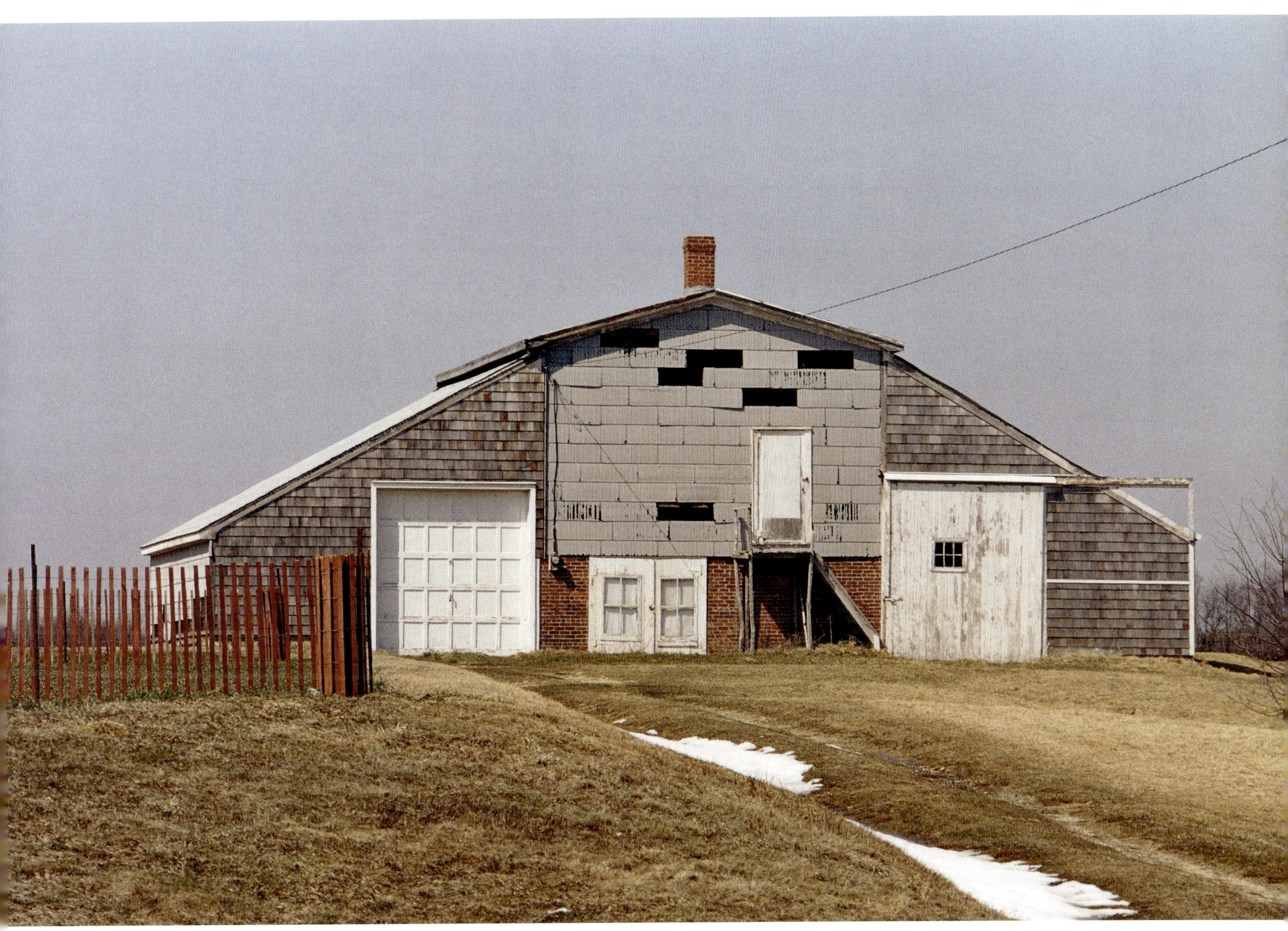

Riverhead

Setting the exact date on a barn is not simple. It is common to be told the date for a house followed by "The barn was probably built at around the same time."

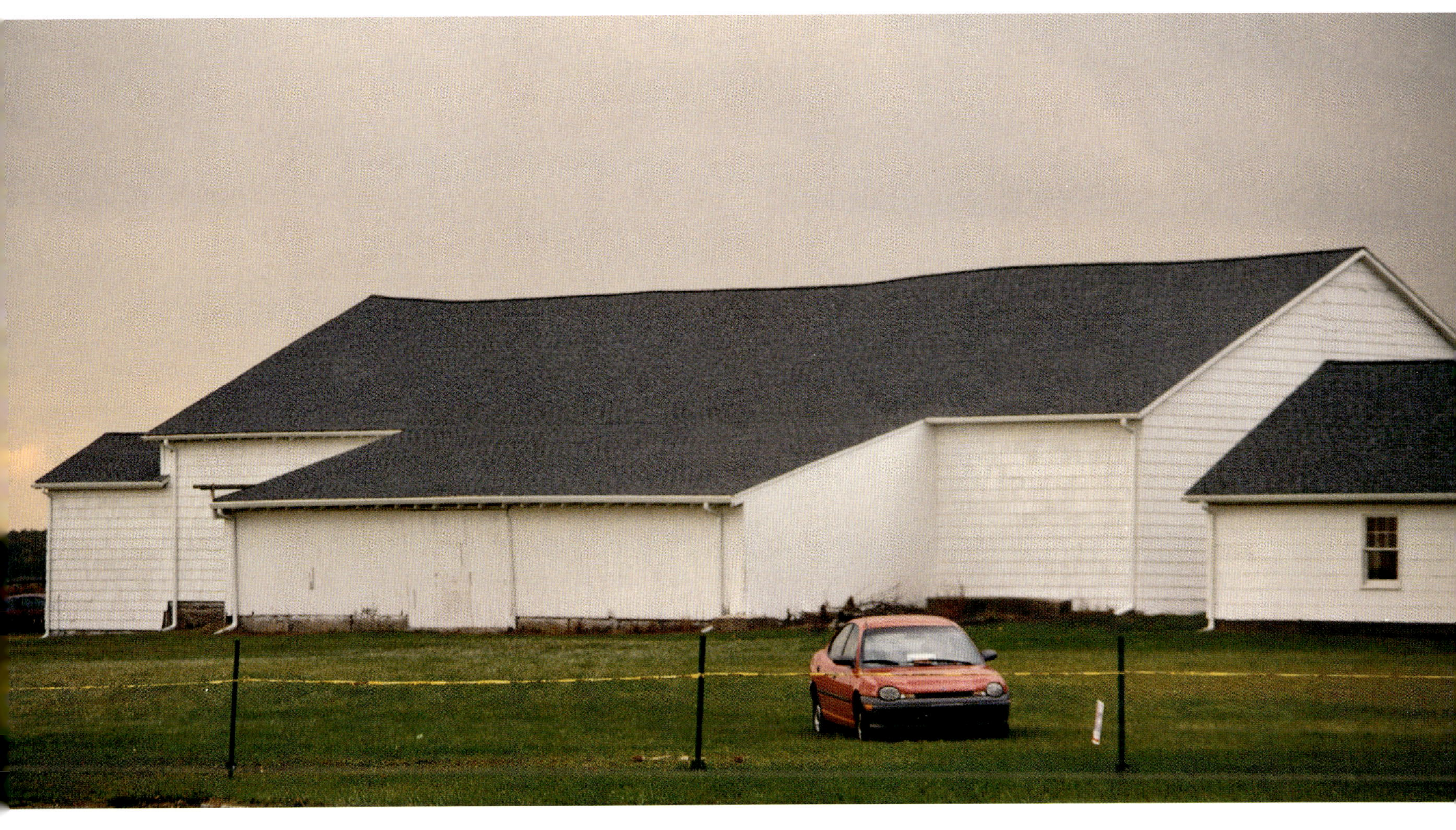

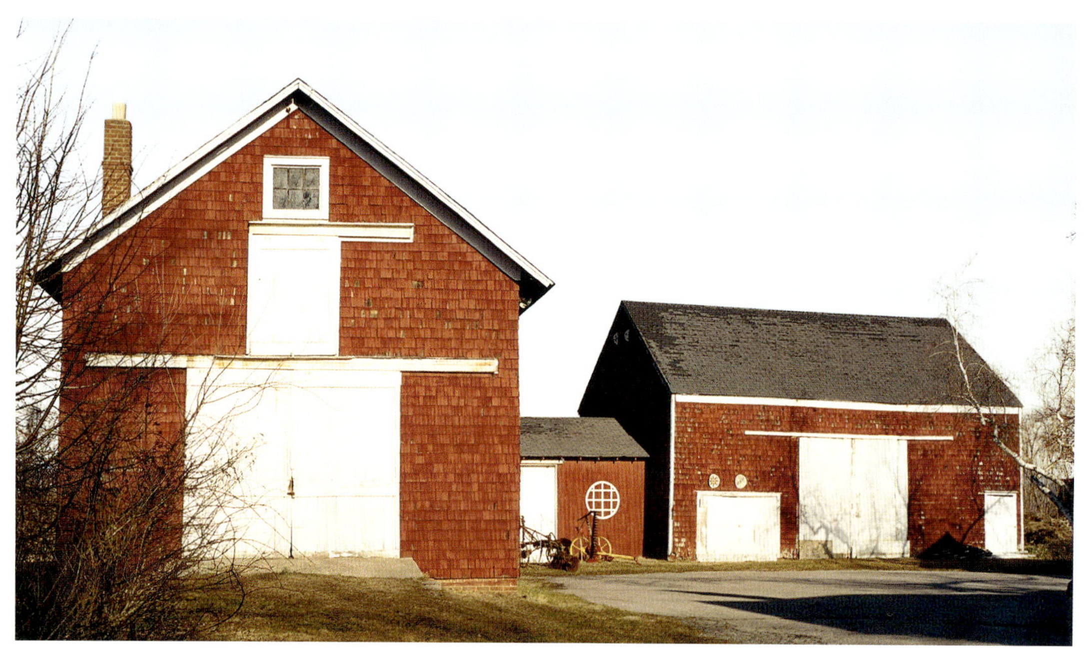

Southold

The eastern portion of the double barn was built in the eighteenth century. The house may be older than the barn and contains timbers that could have come from a barn frame.

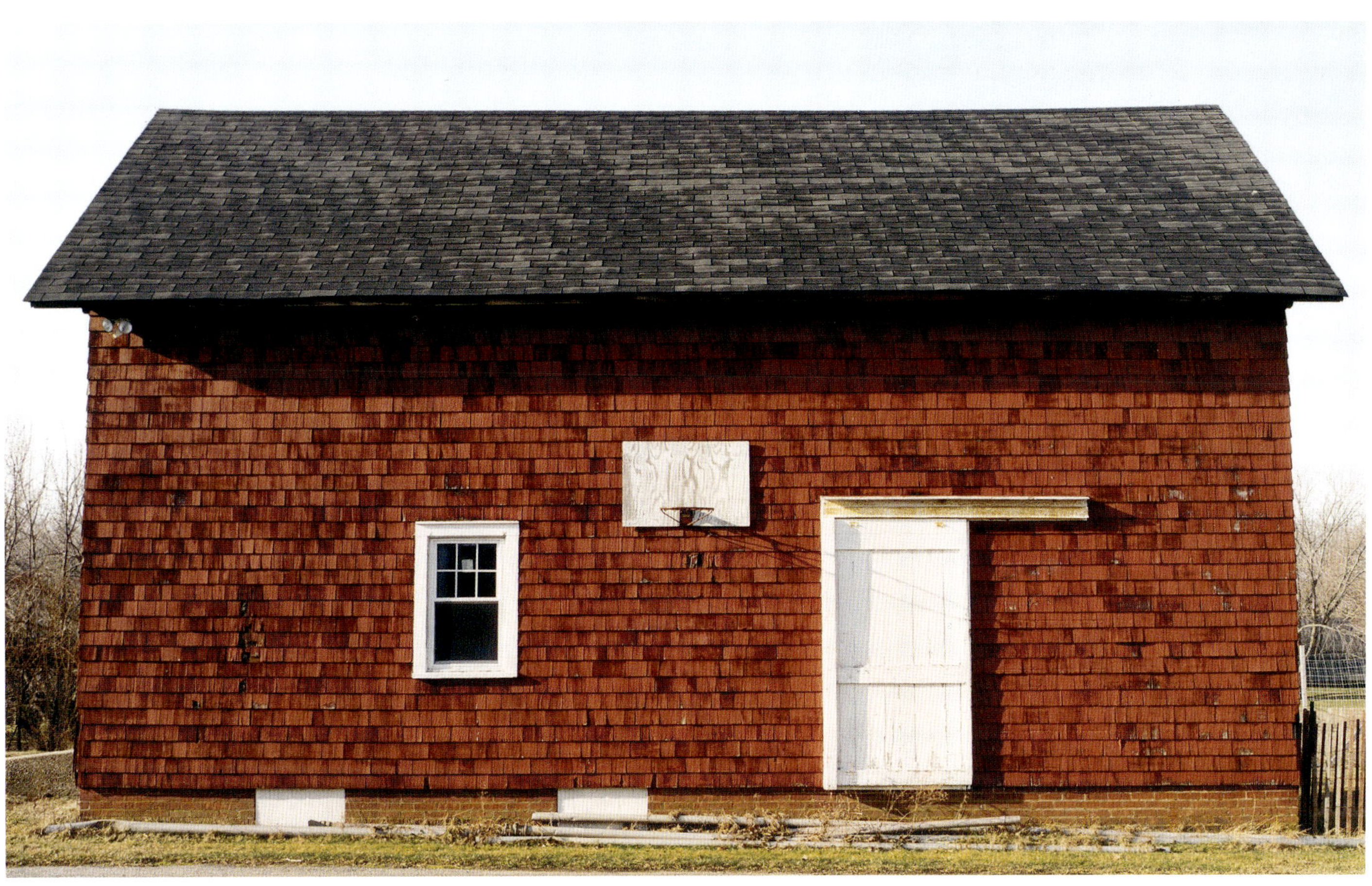

Mattituck

Stables account for 6 percent of the barns in the Southold survey.

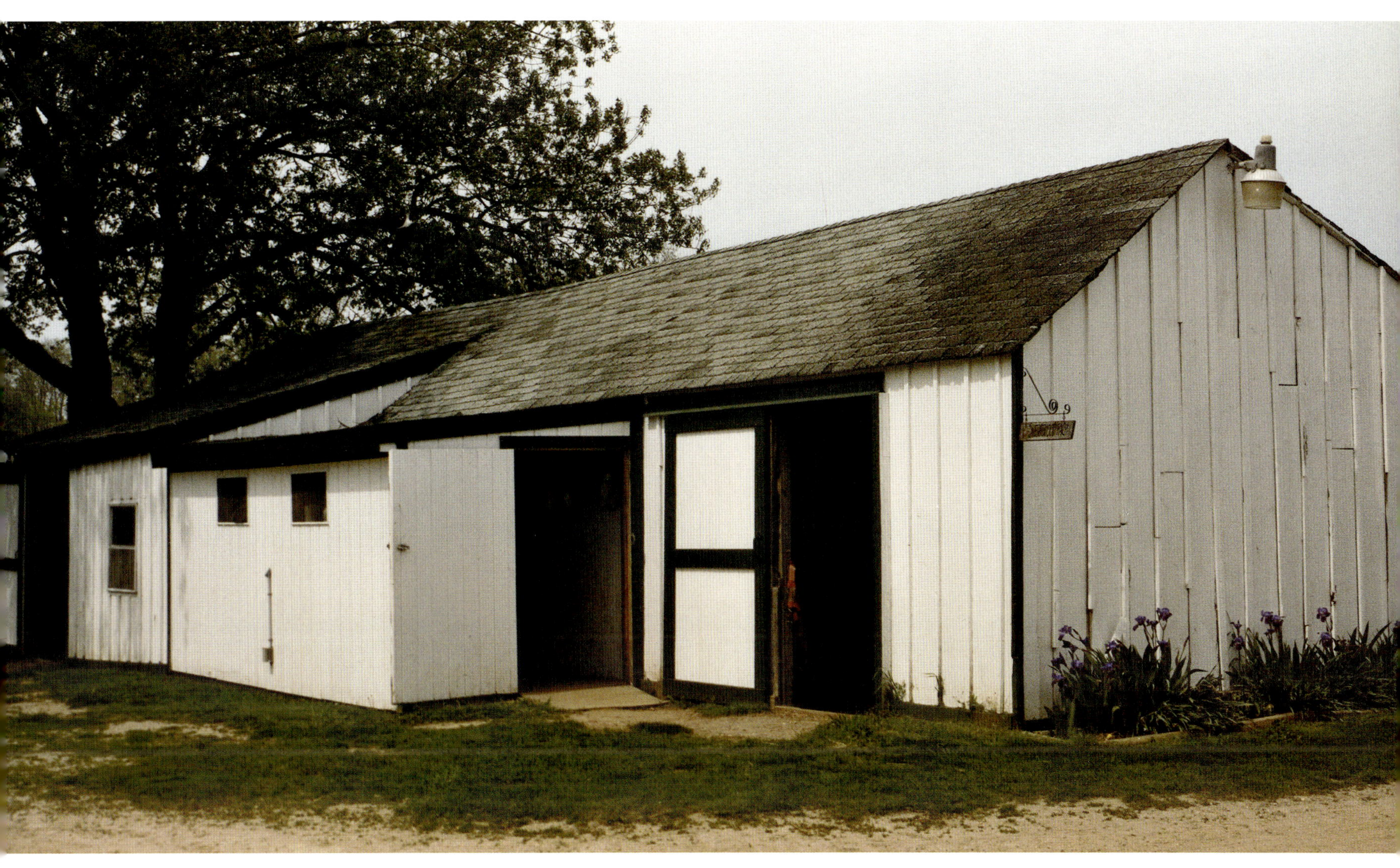

Peconic

A metal pole barn was recently attached to the west side of this interesting barn.

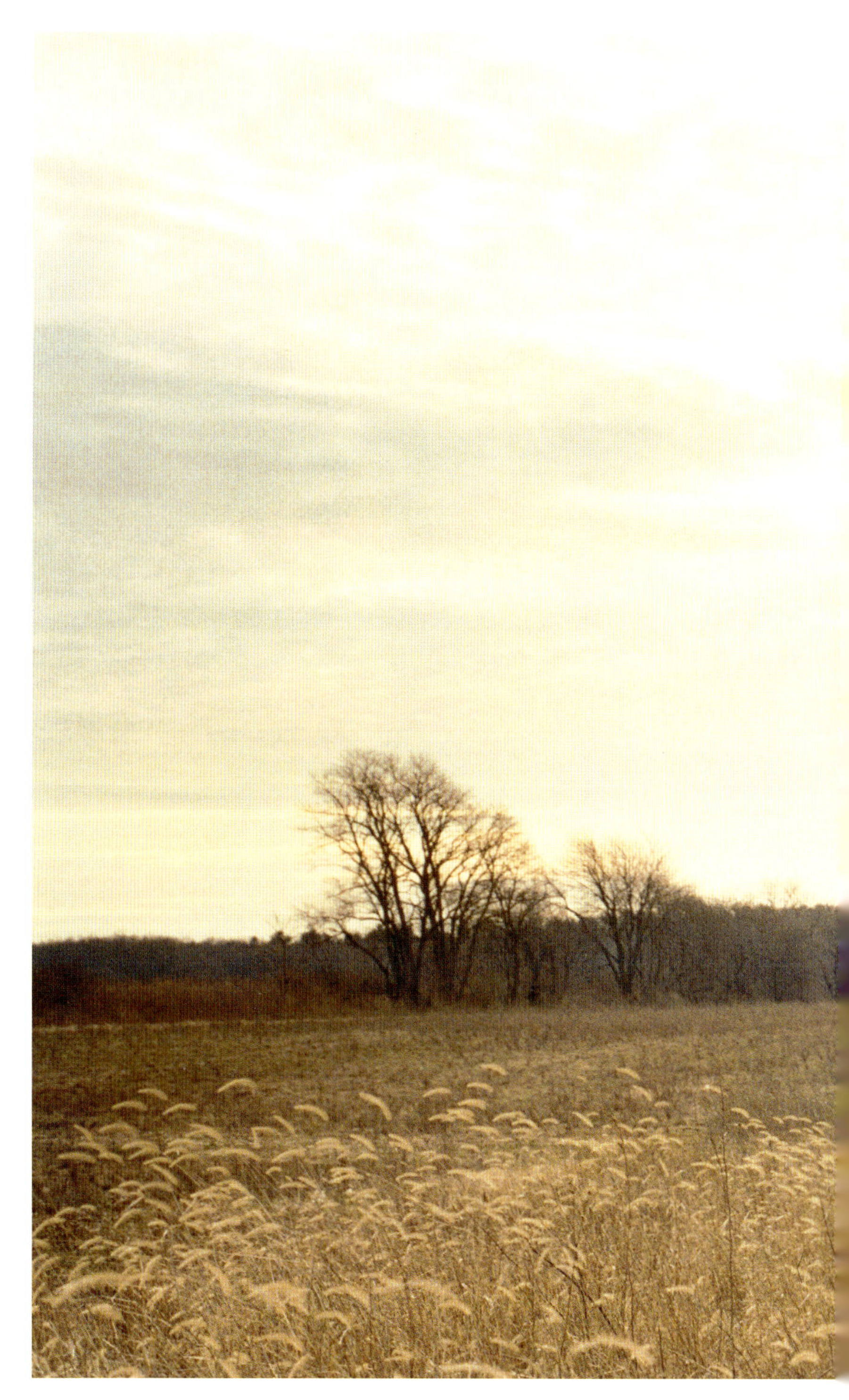

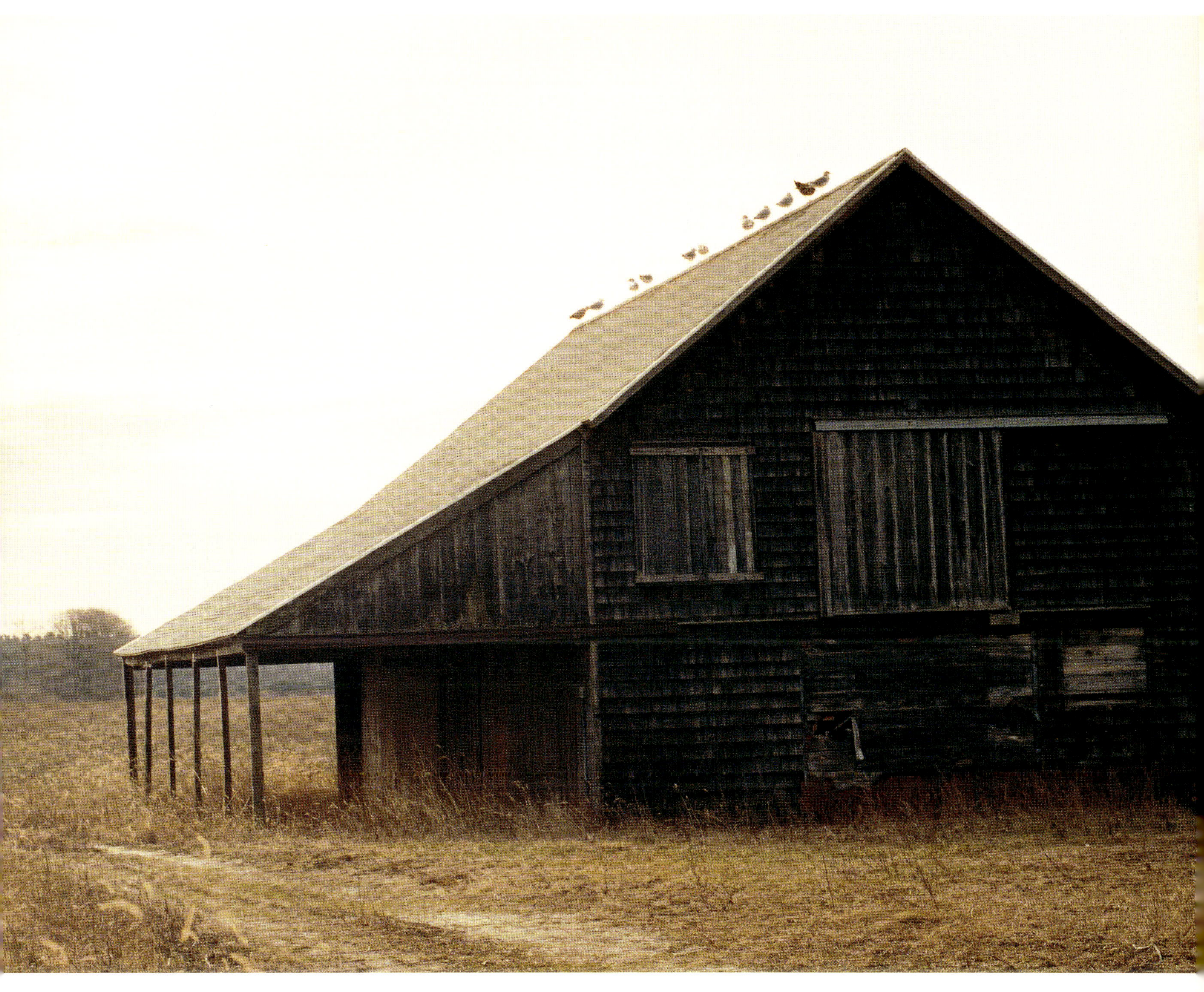

Cutchogue

Barns are adapted for use as dwellings and for business.

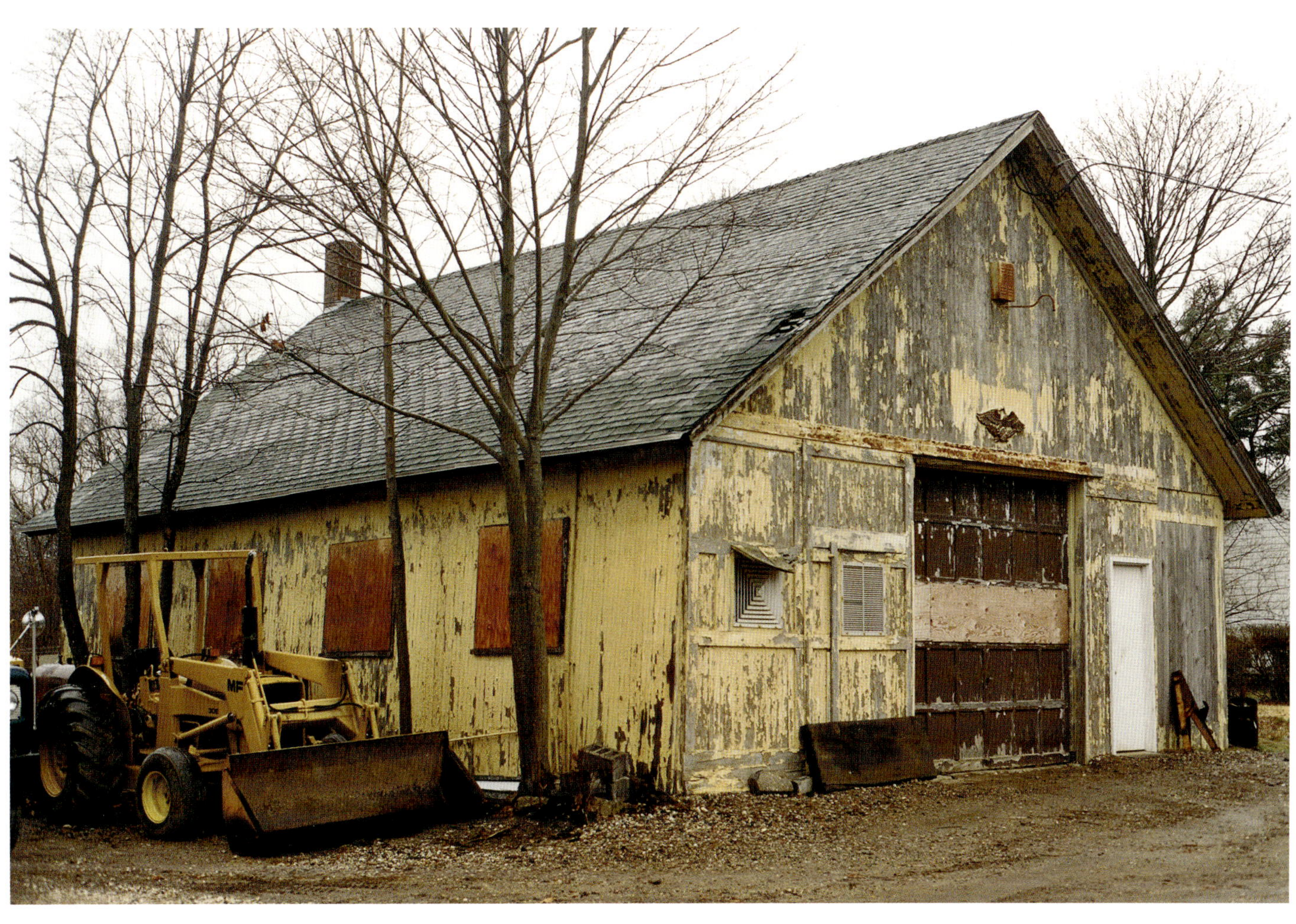
MF

Cutchogue
Collections of farm outbuildings are both common and visually interesting.

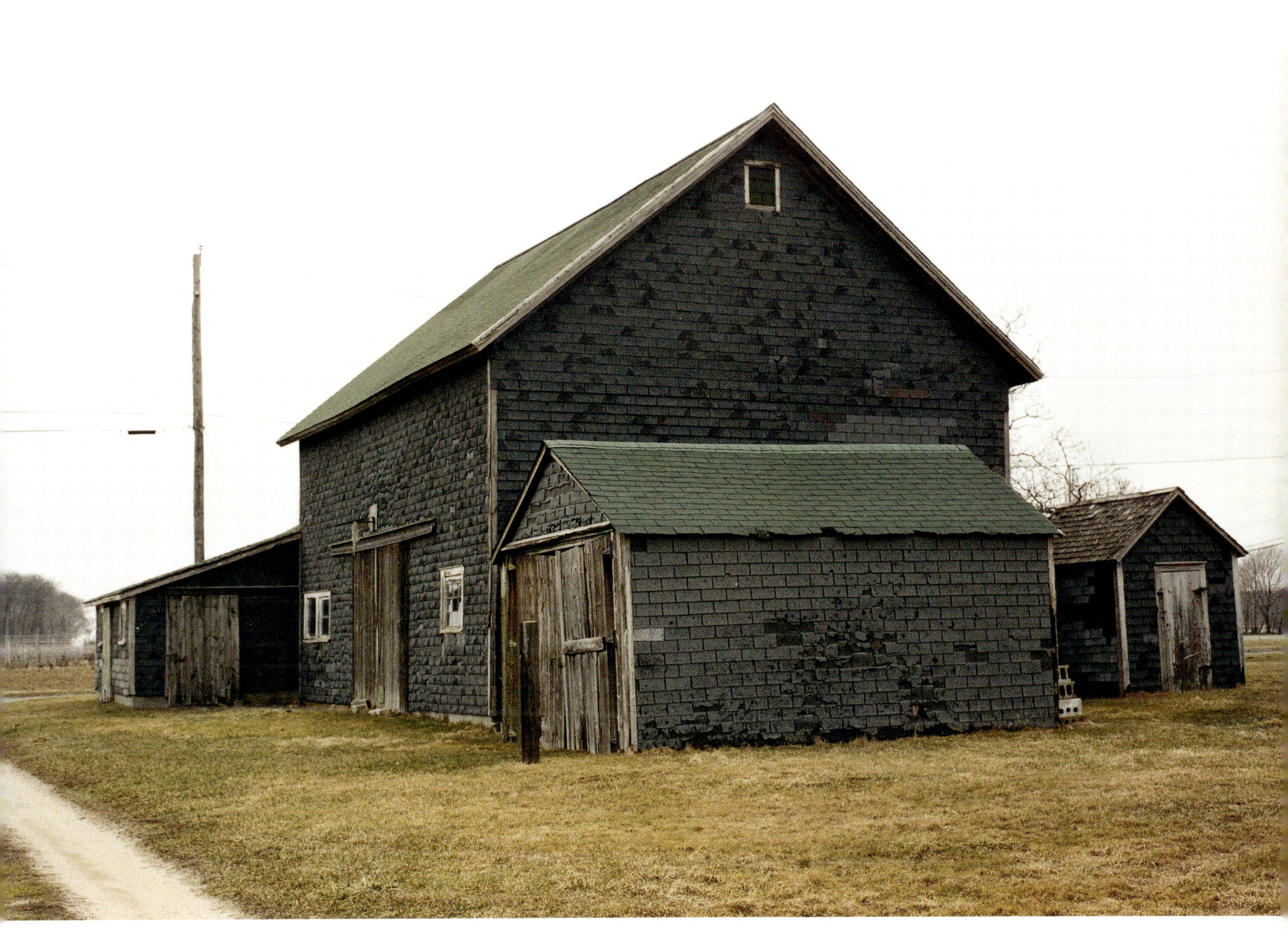

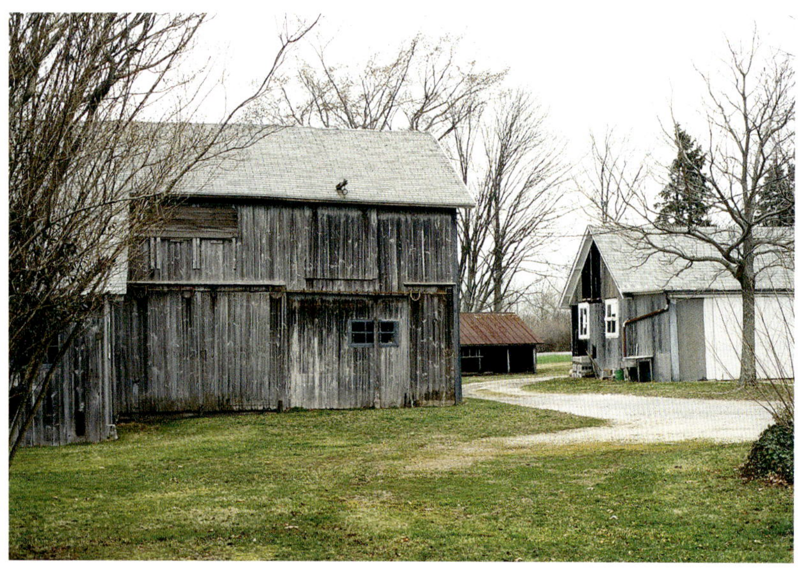

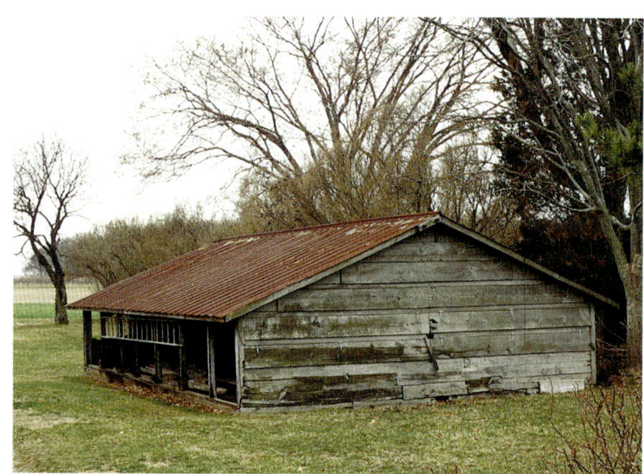

Cutchogue

The asphalt-shingled section of this double barn was built in the early eighteenth century. The vertical plank barn was added in 1910. There is a large shed addition (1890) on the early barn.

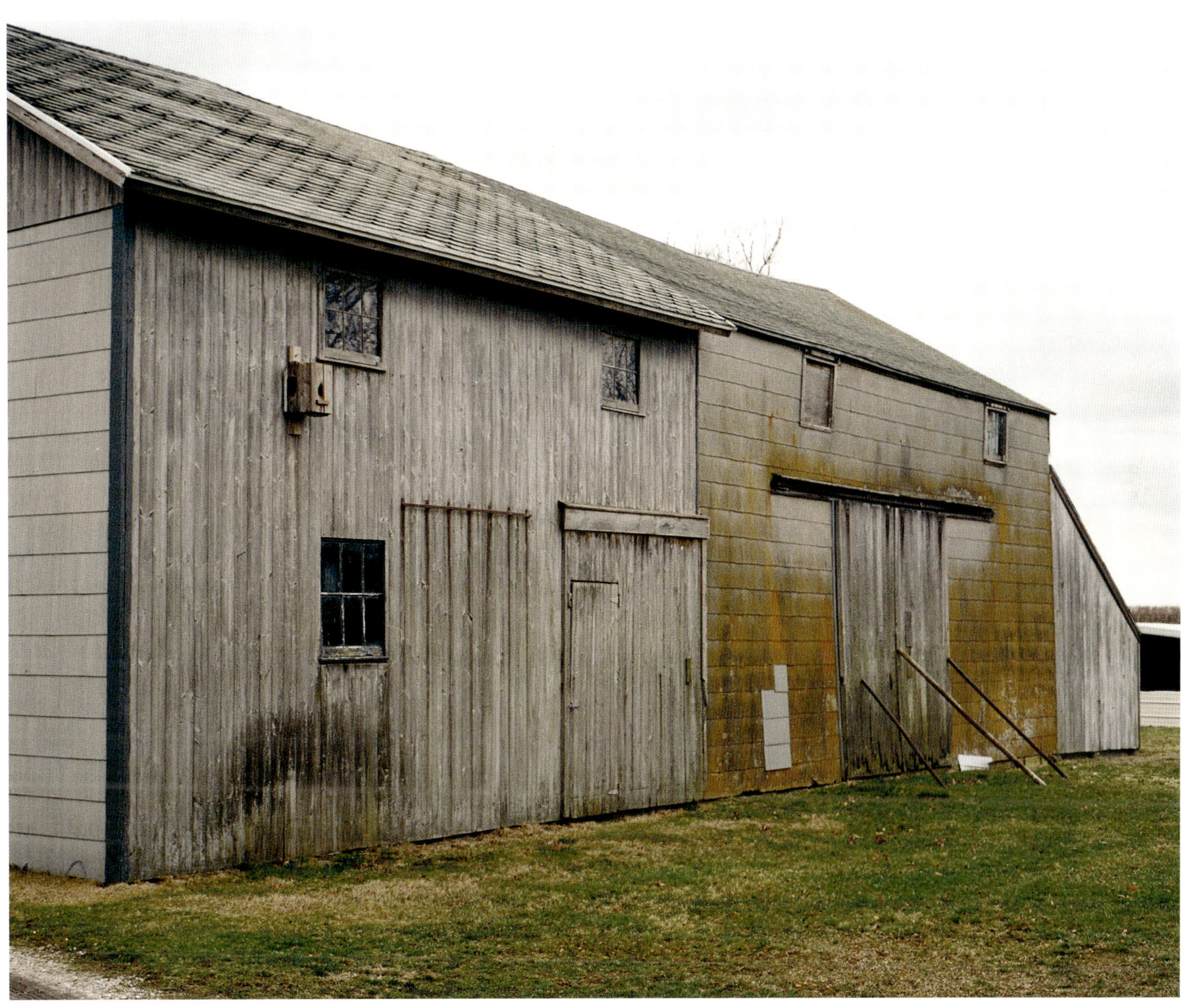

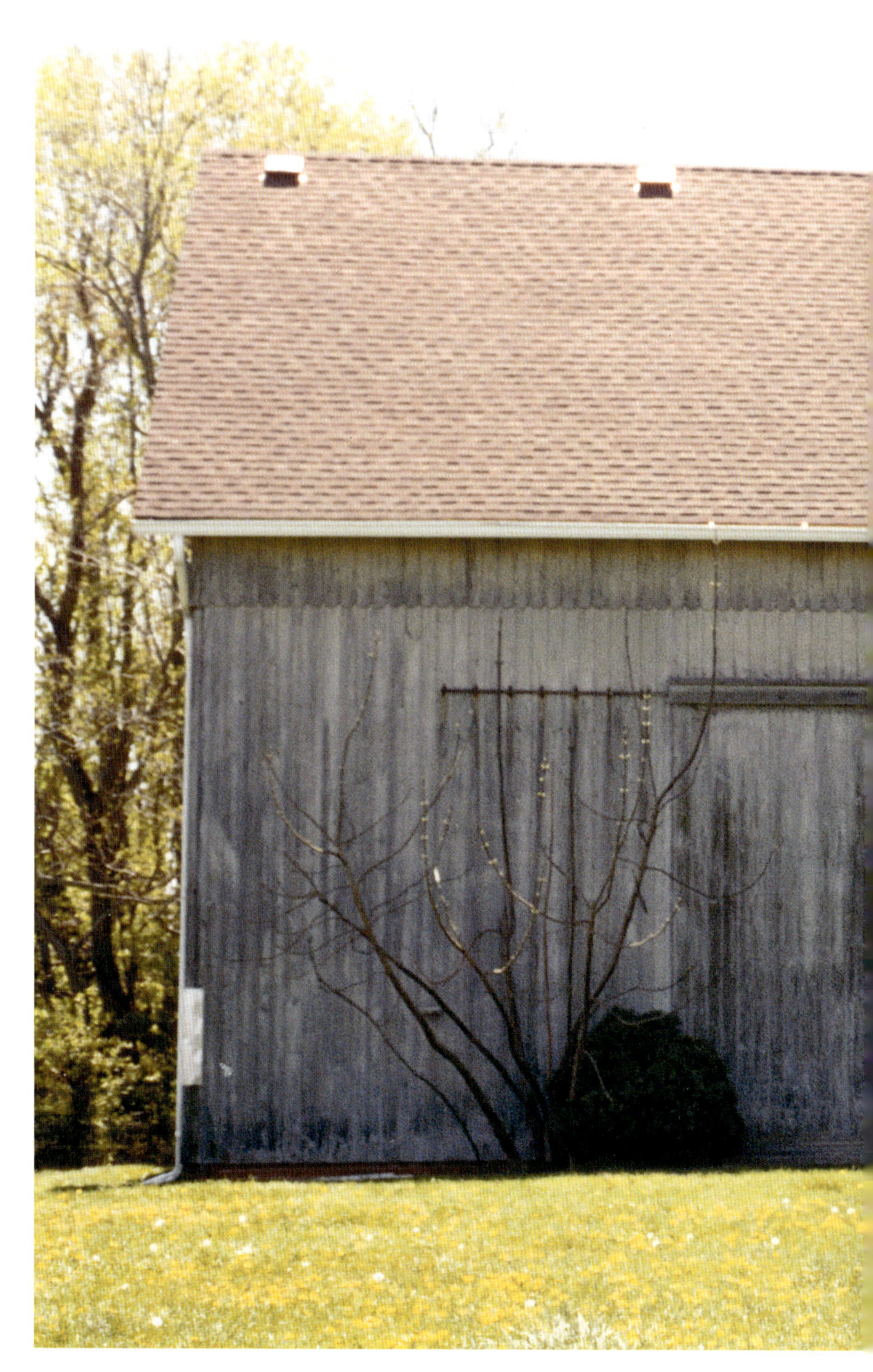

Oregon
The main barn was built in 1860.

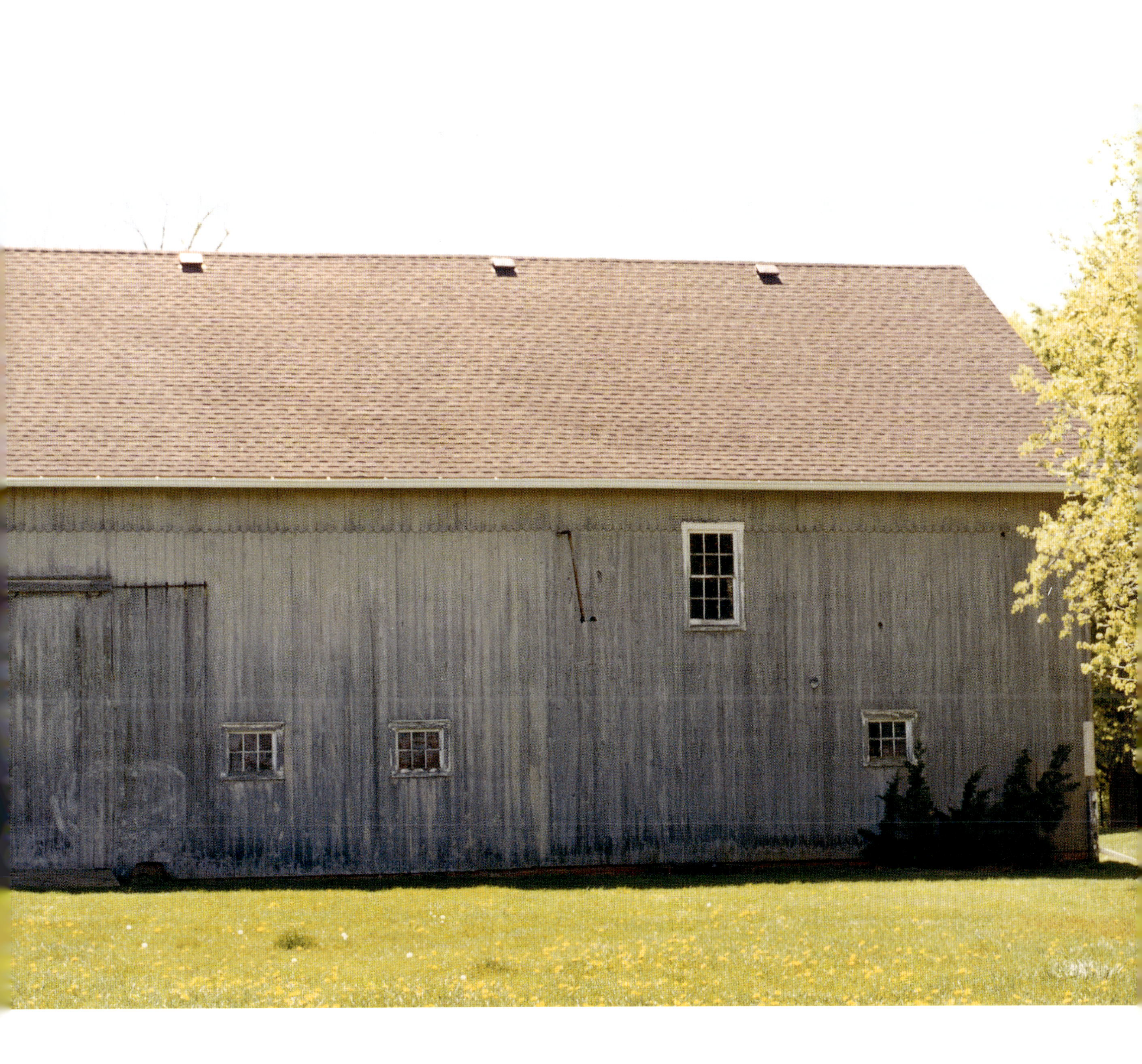

Southold

A barn in its agricultural setting evokes the feelings that we hold about our agrarian past.

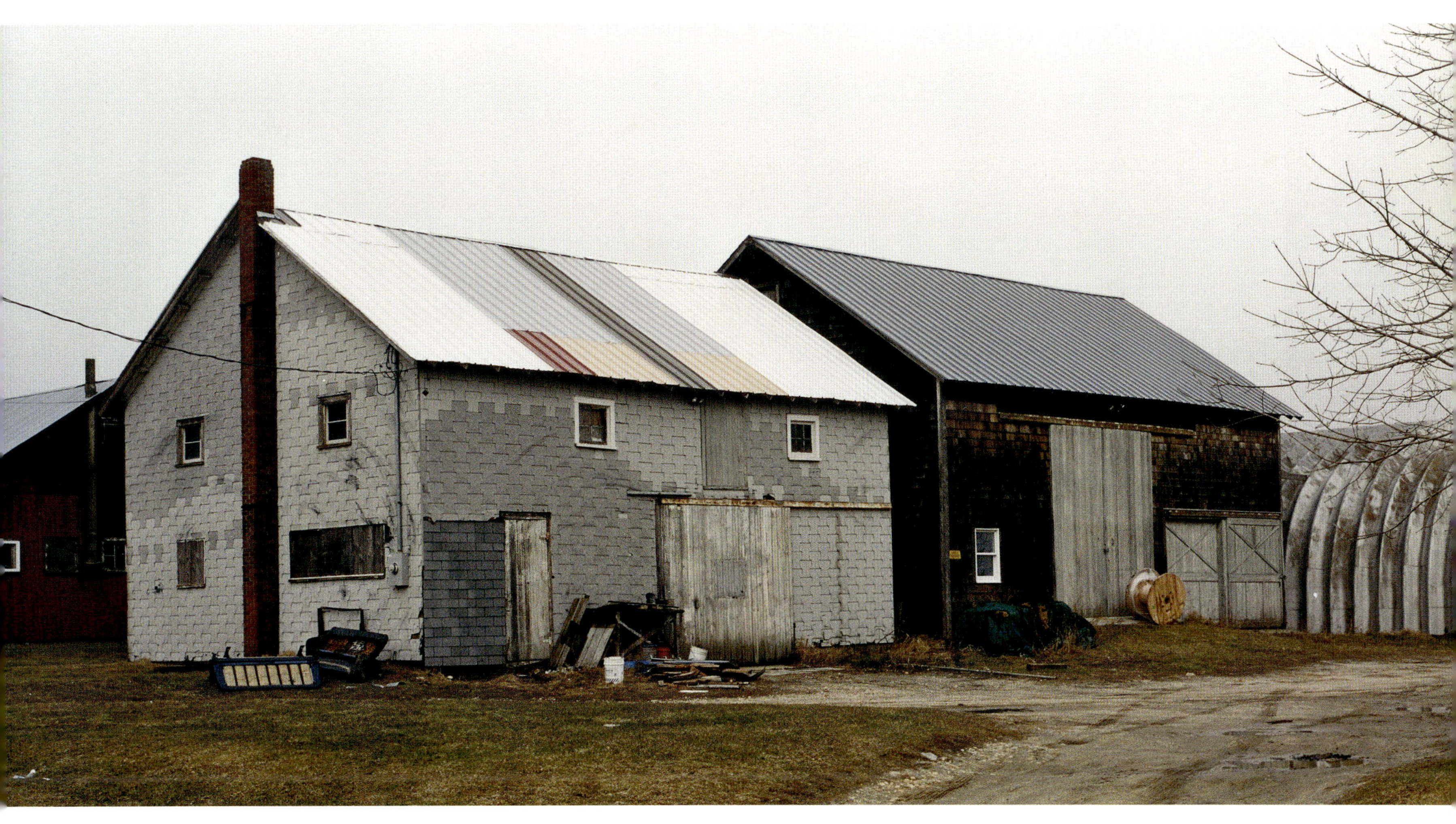

Riverhead

This photo is twenty years old.
The barn is still standing. The bike is gone.

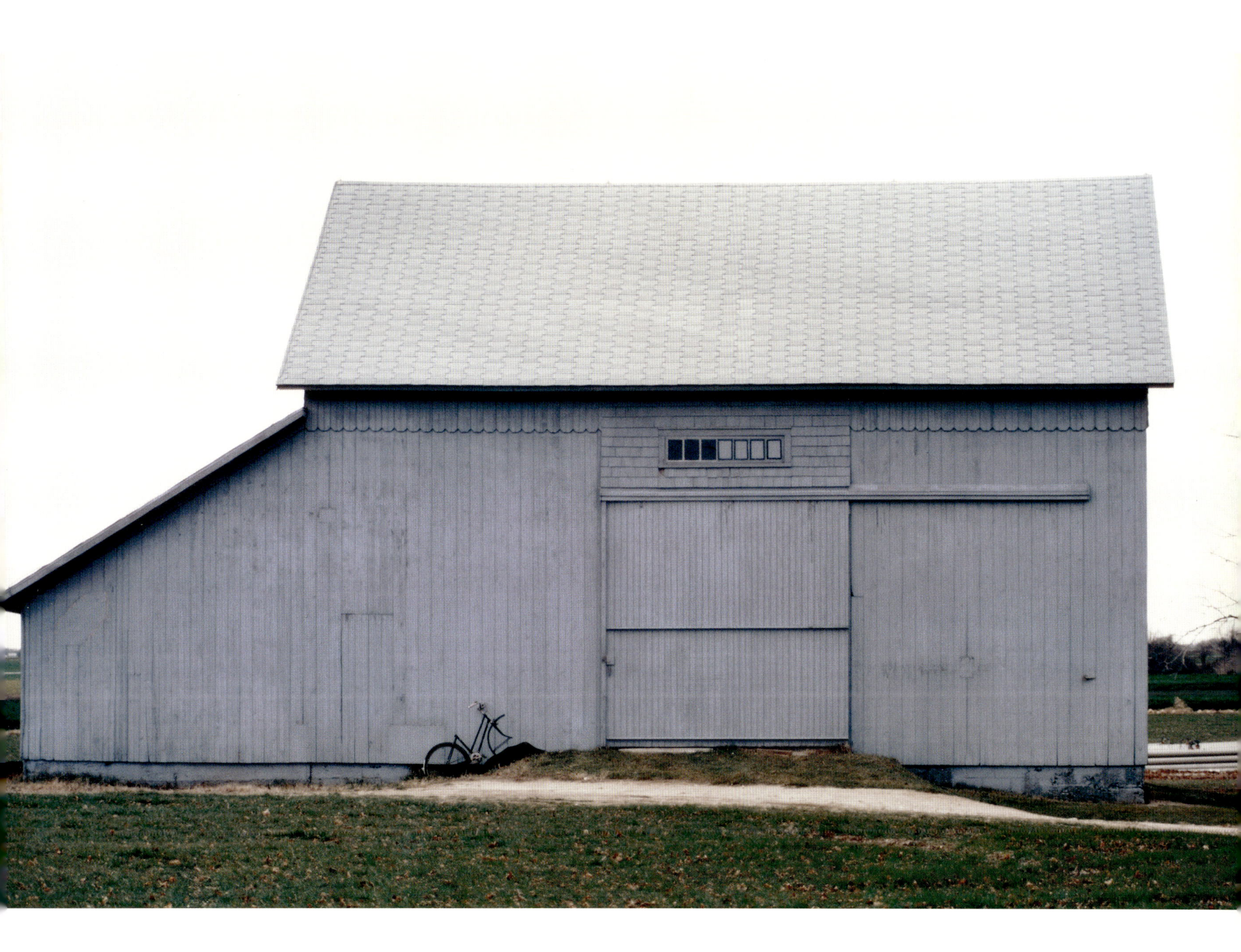

Peconic

Quonset hut barns appeared after World War II. All of the structures on this site were brought from Jericho in the early 1980s.

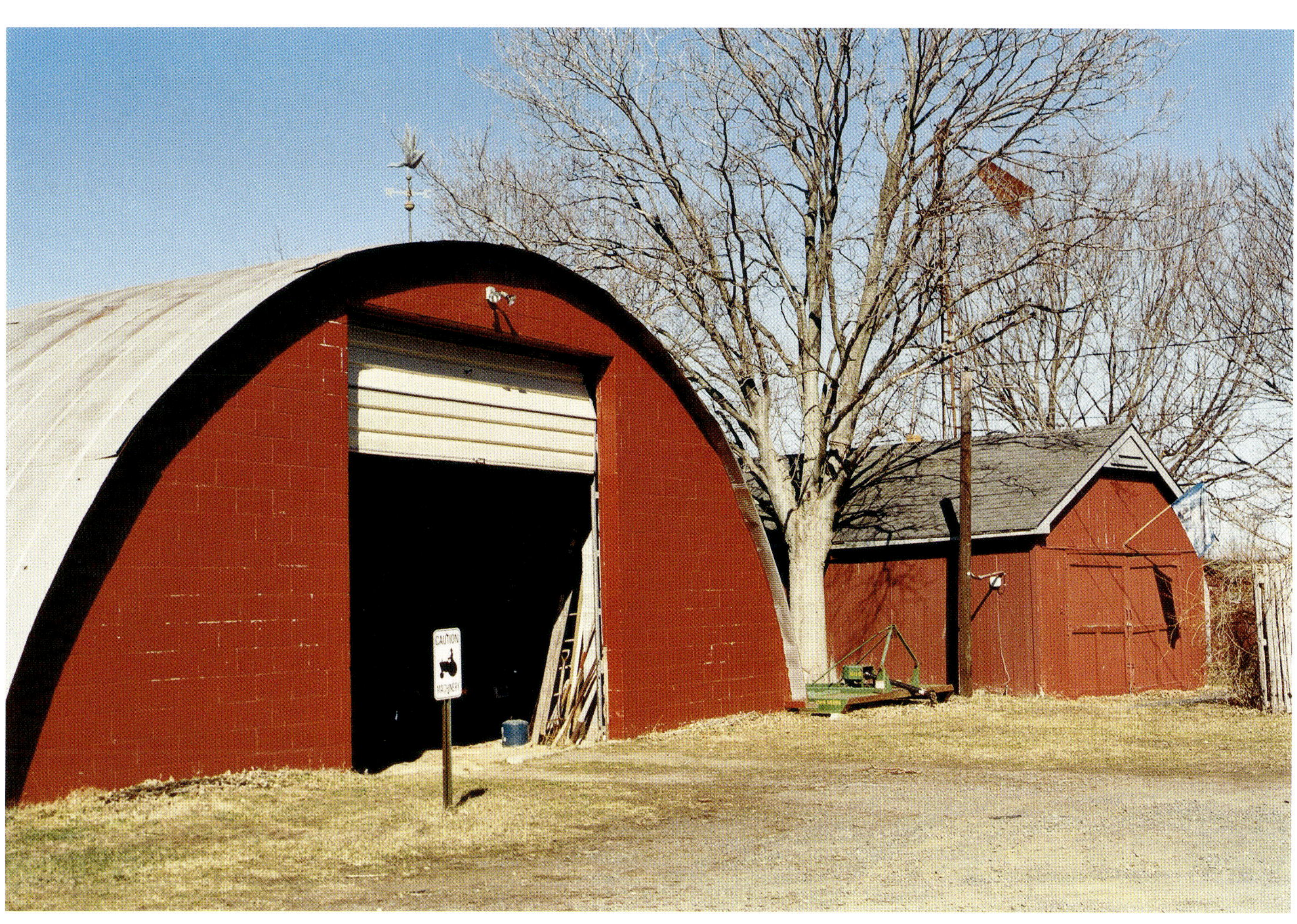
CAUTION

Southold

There are two additional nineteenth-century, formerly attached, wood-shingled barns on the property. This smaller barn was built in the 1930s.

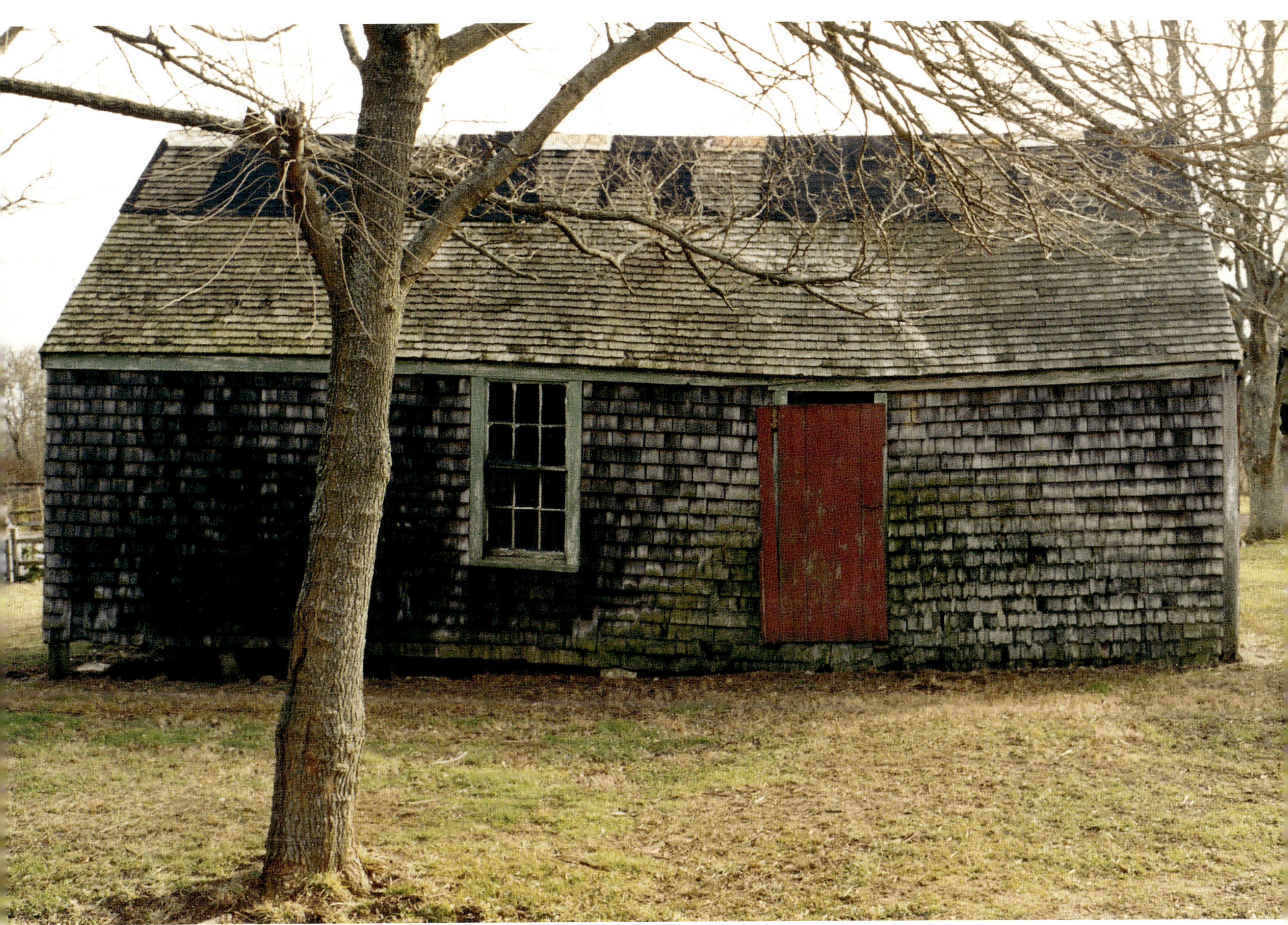

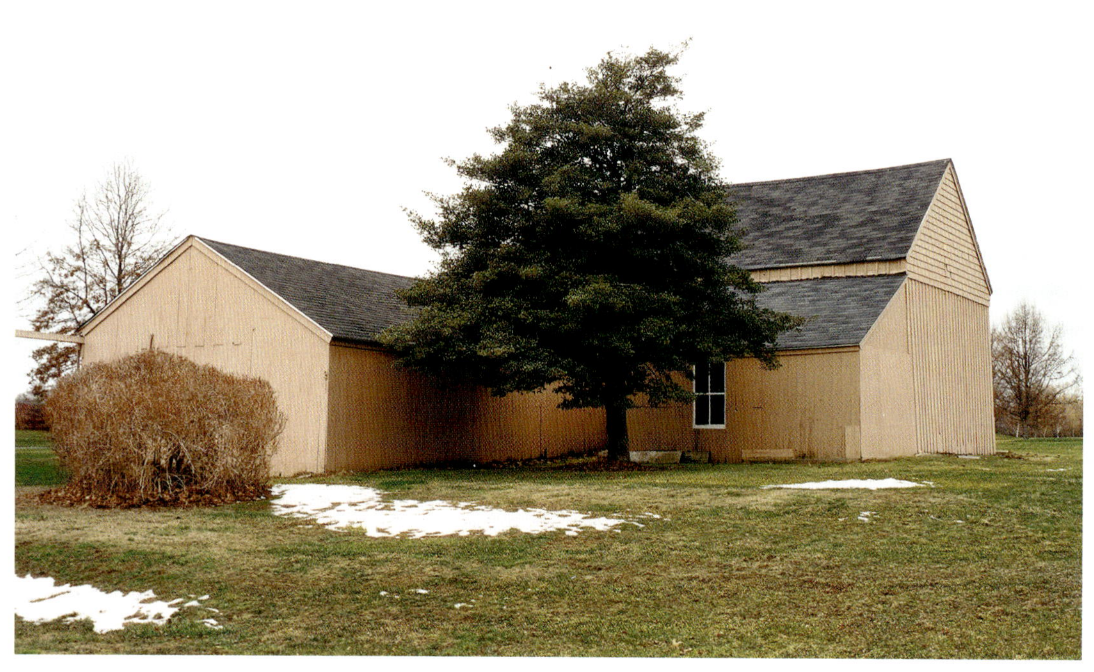

Greenport
Barns are the best-known examples of vernacular architecture.

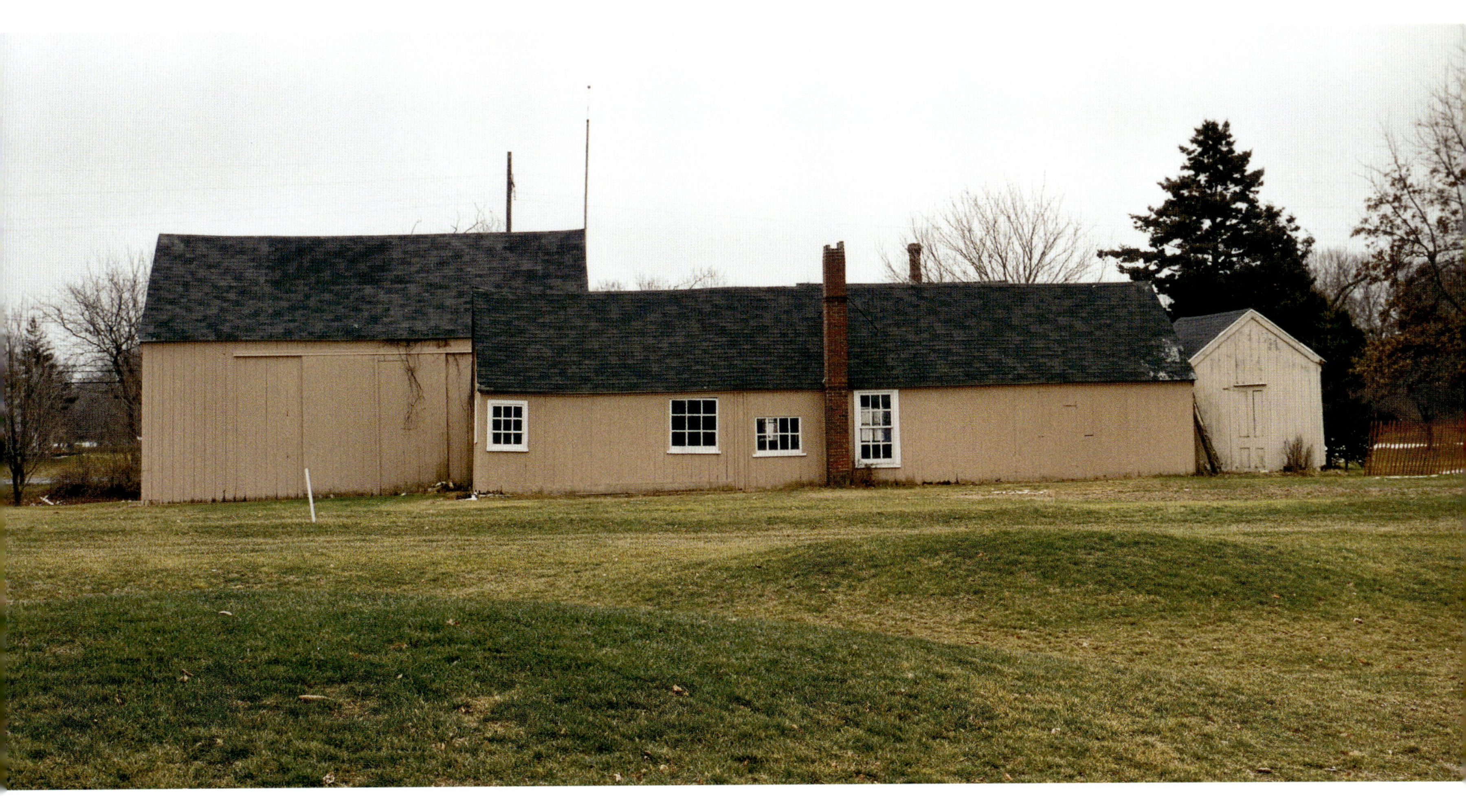

Mattituck

The buildings were constructed recently for vineyard use.

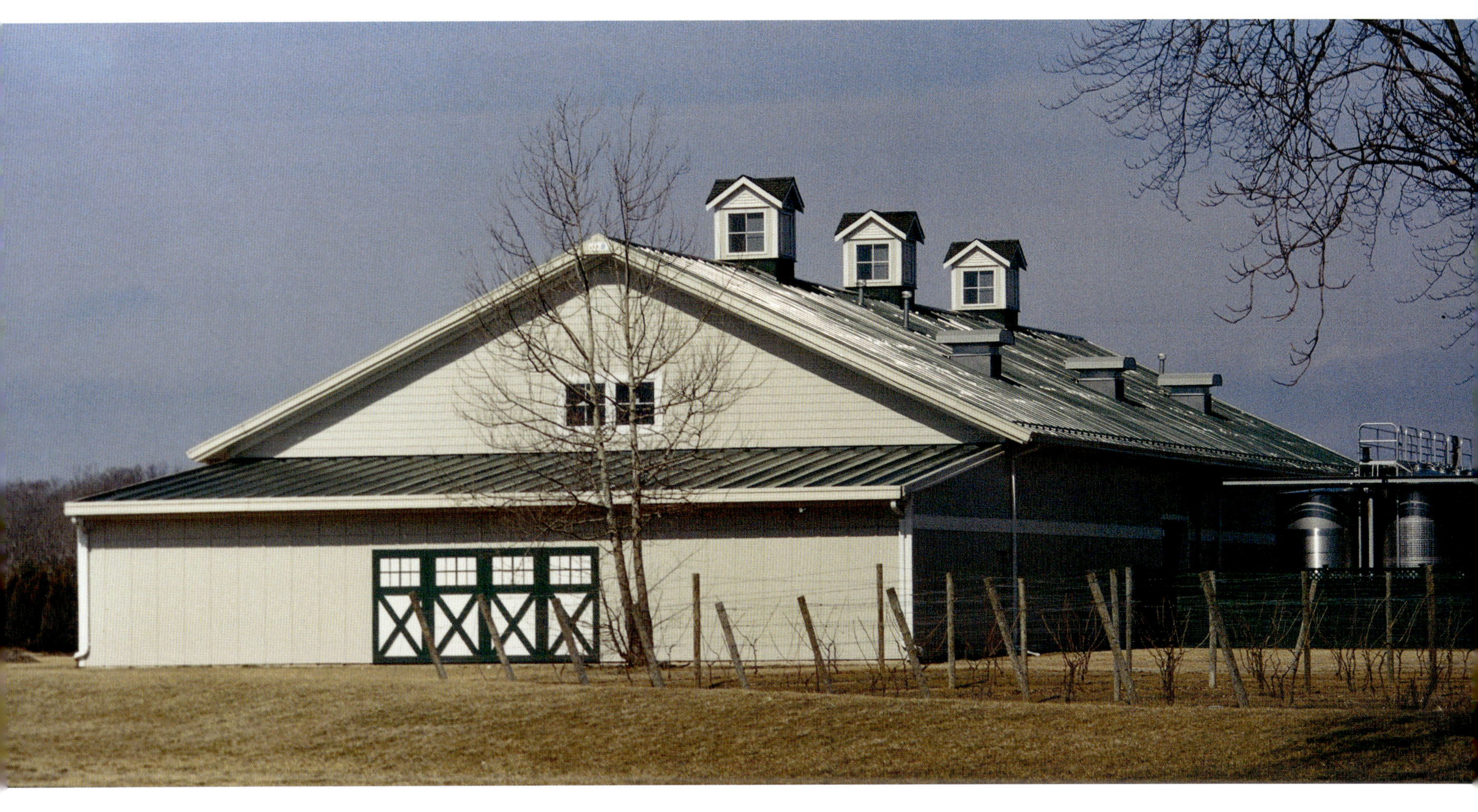

Peconic
The fate of some barns is unclear.

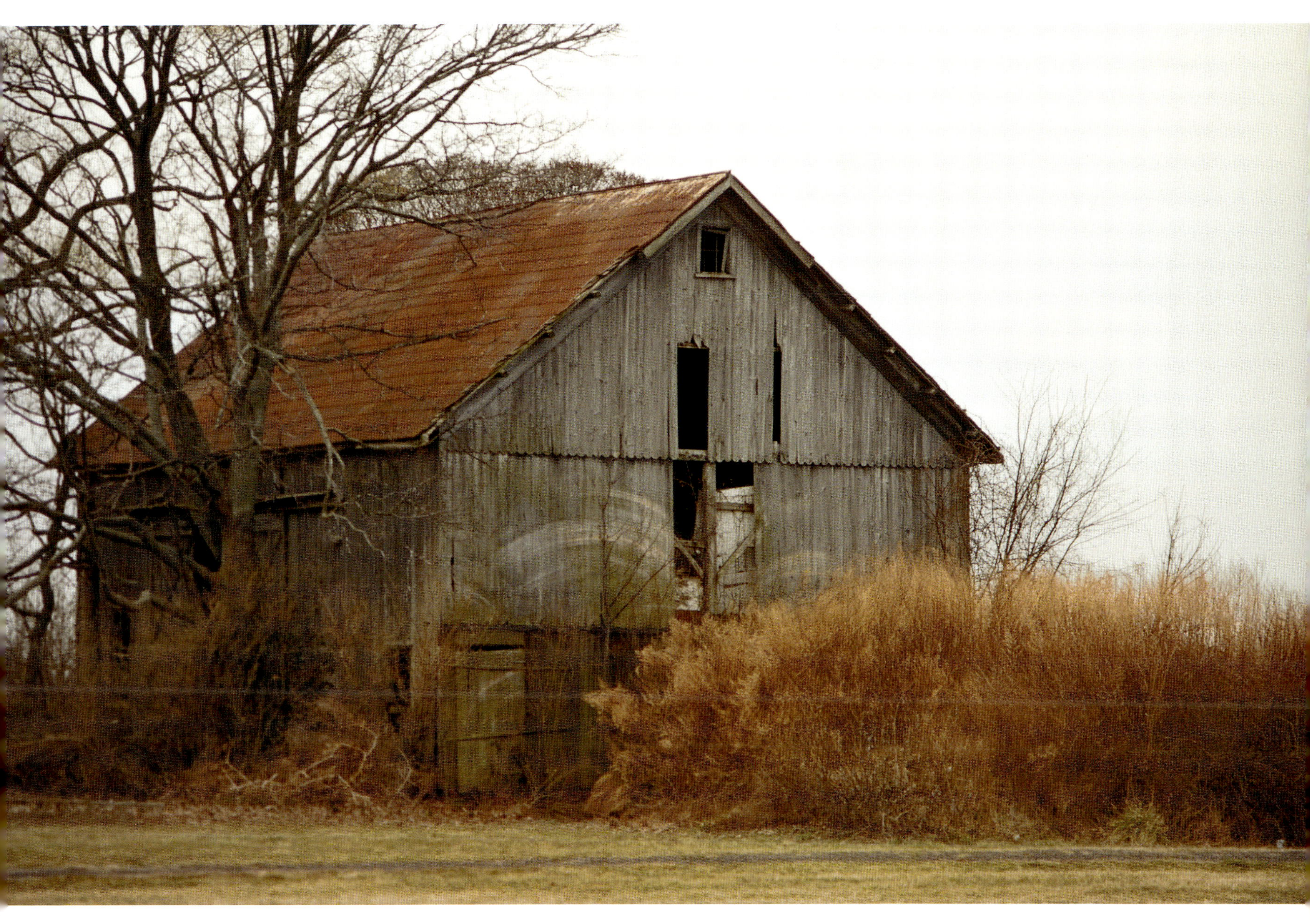

Peconic

Pole barns are relatively inexpensive, practical, and universal.

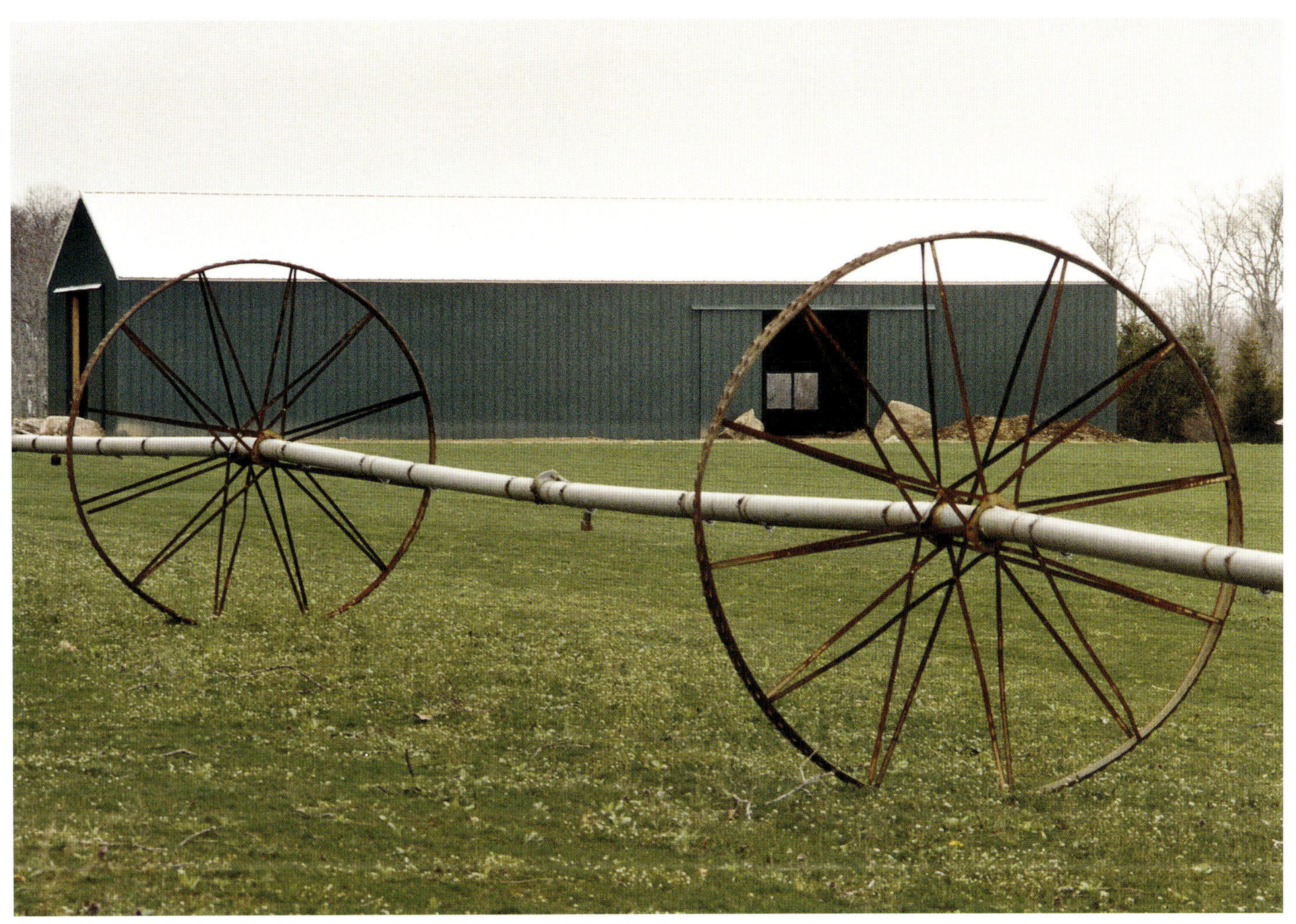

Bayview

This barn, built in Greenport "off Carpenter Street," was facing demolition. The present owner asked if he could have it. He moved it to Bayview.

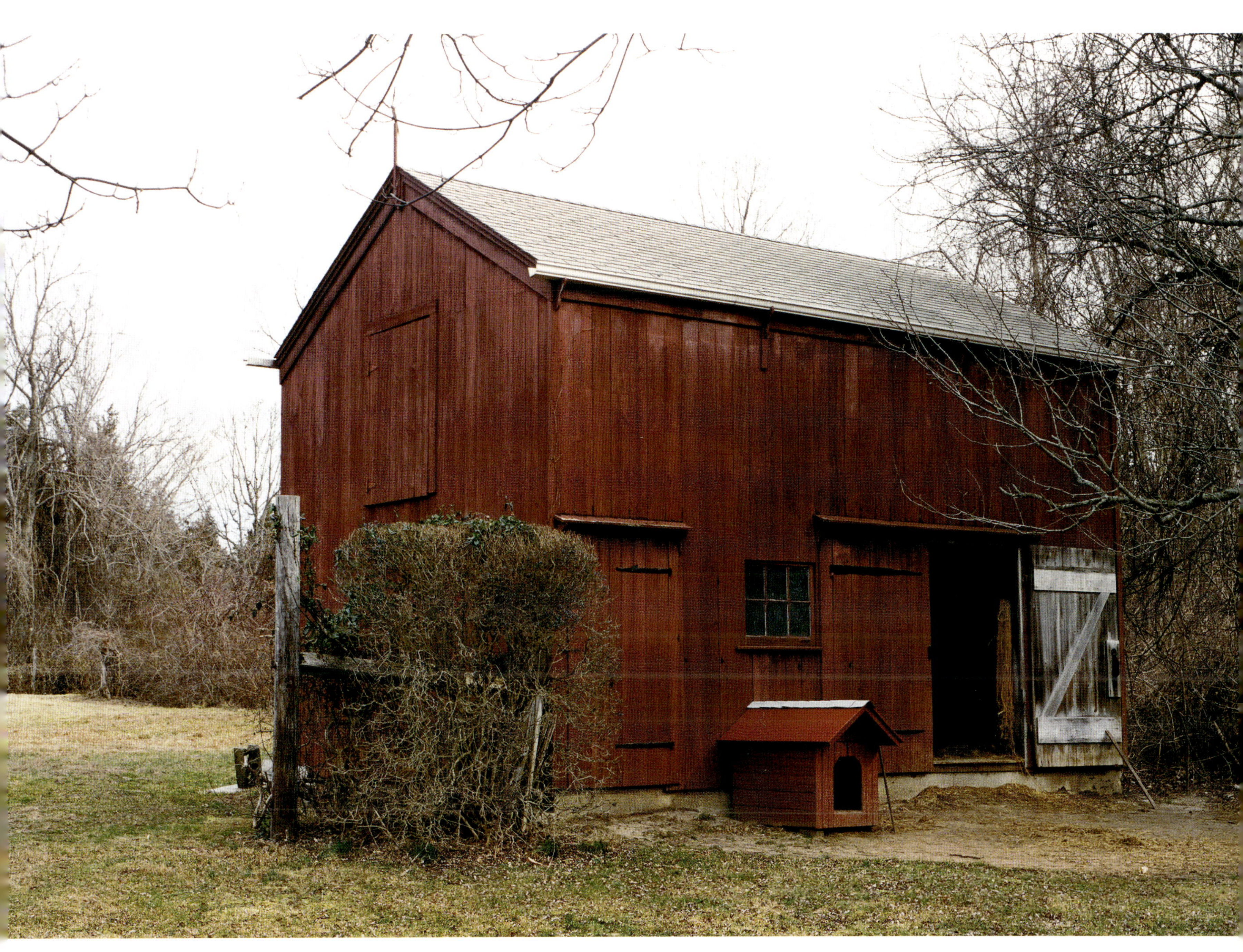

East Marion
Small brick chimneys are common alongside barns.

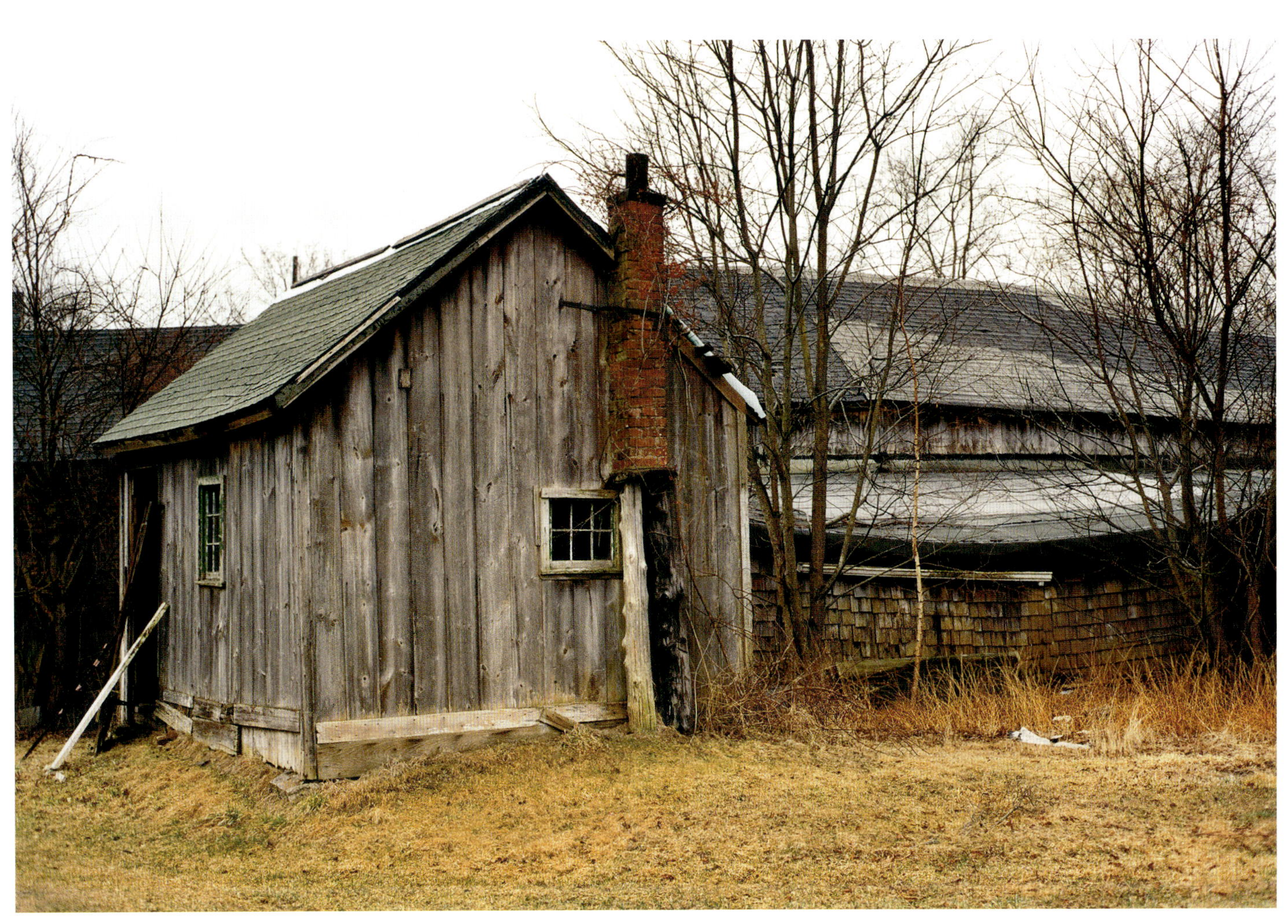

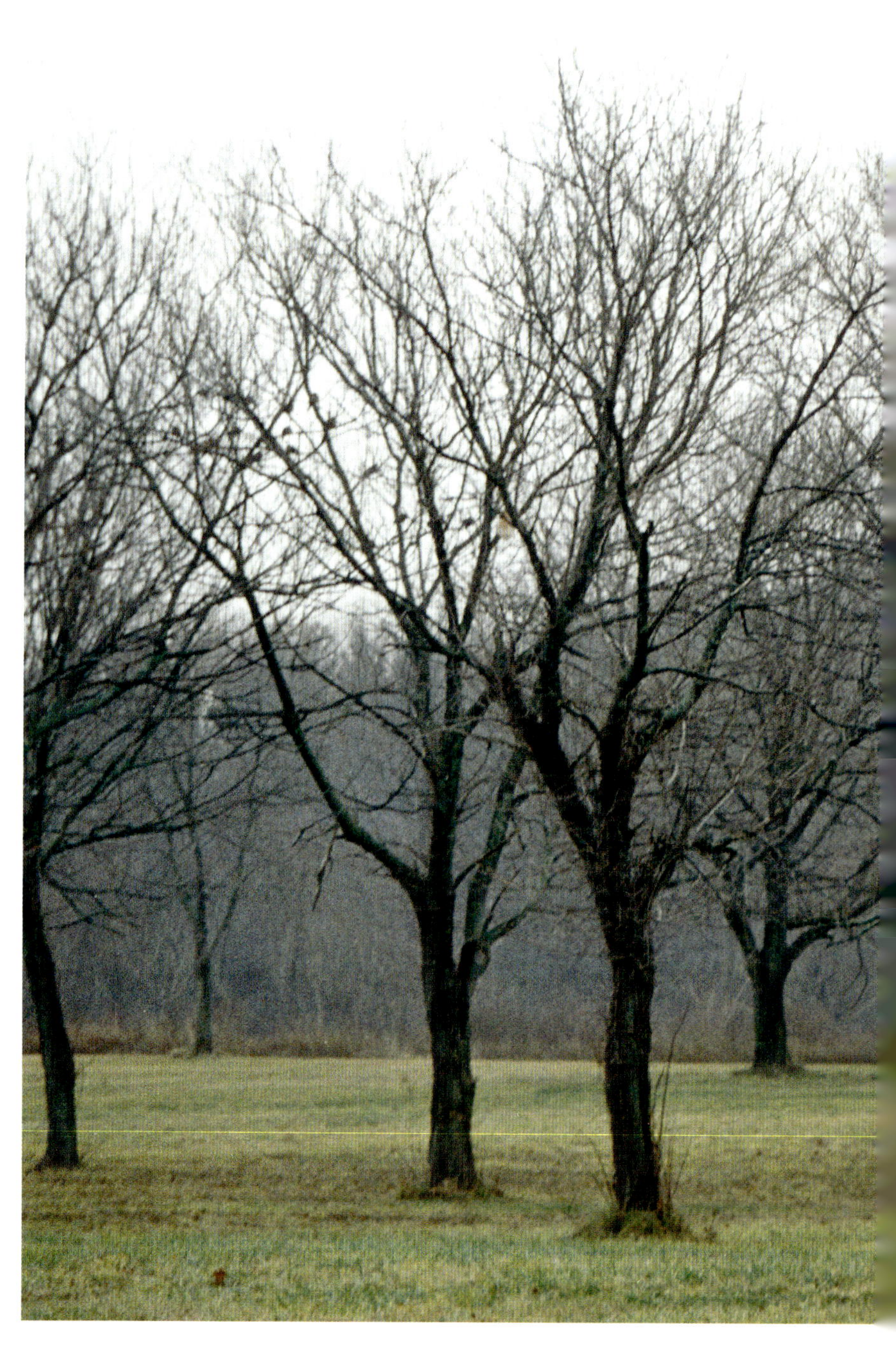

Southold

When we envision barns, do we see the structure or do we see the structure in its setting?

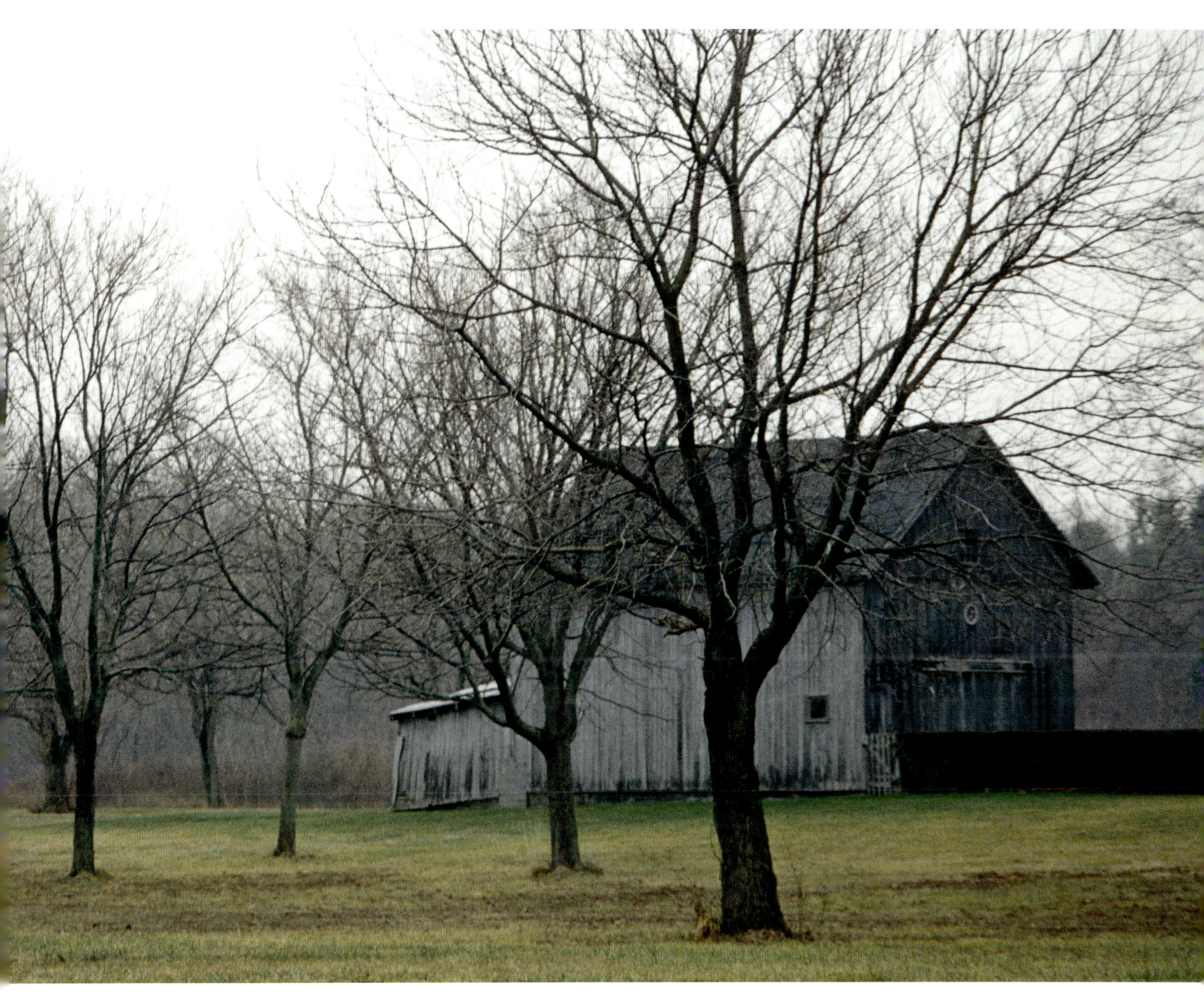

Greenport

This barn is more than one hundred years old.
Prior to its current stable use it was a chicken coop.

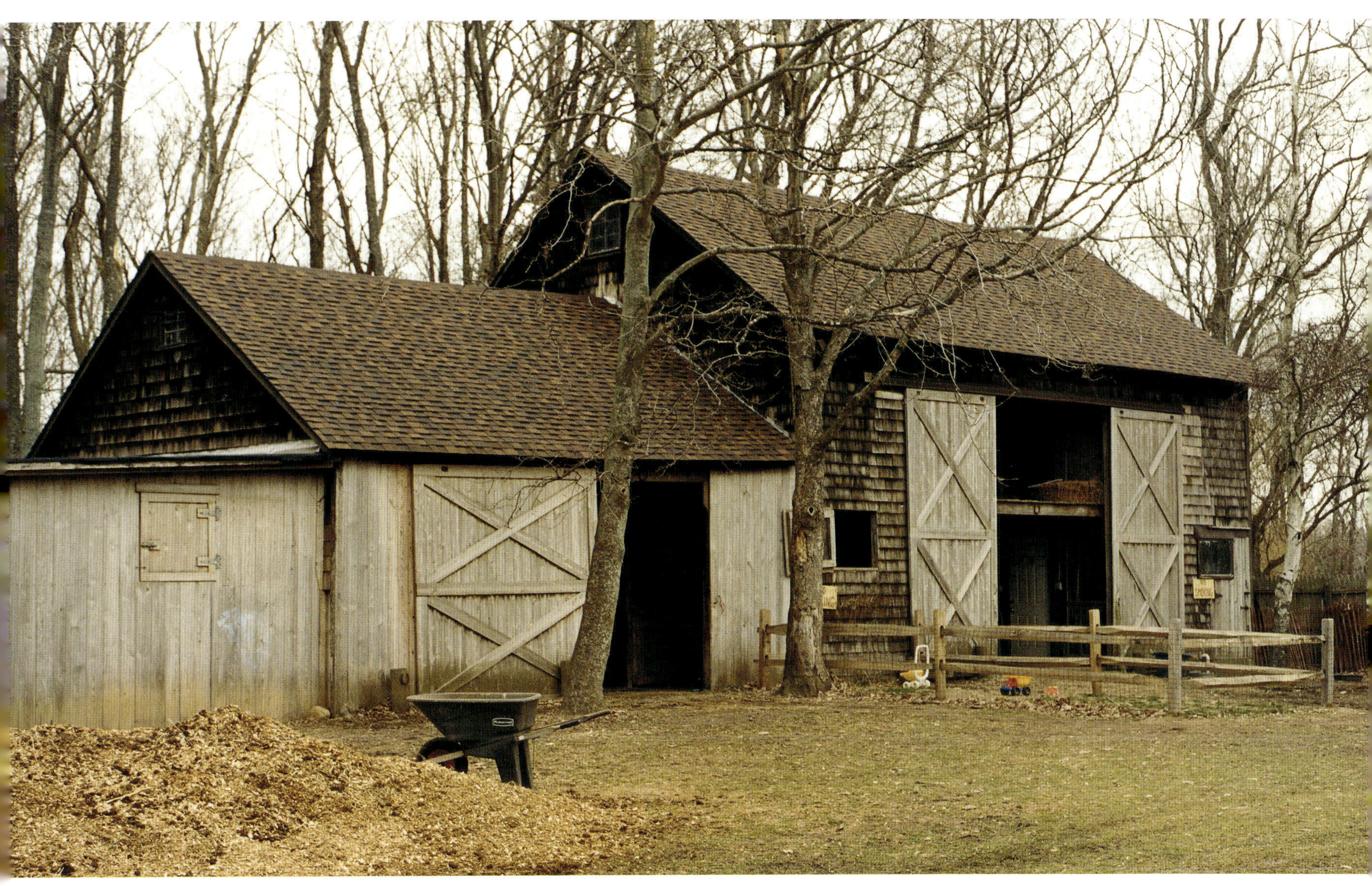

Orient

This picture illustrates the variety of roofing materials used on barns.

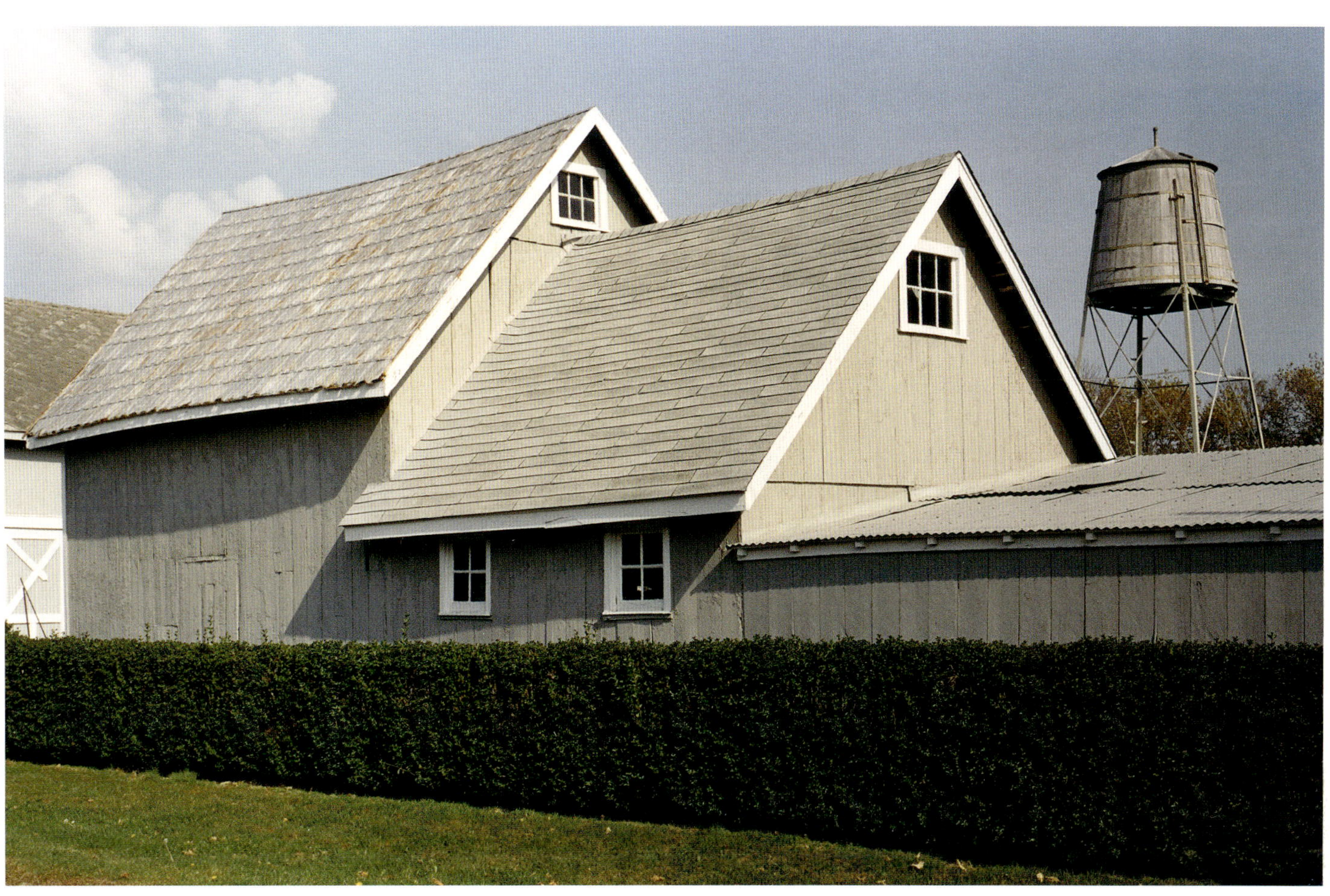

Oregon
One portion of the quadruple barn is an English hay barn (1850–80) that was moved, attached, and altered for potato storage.

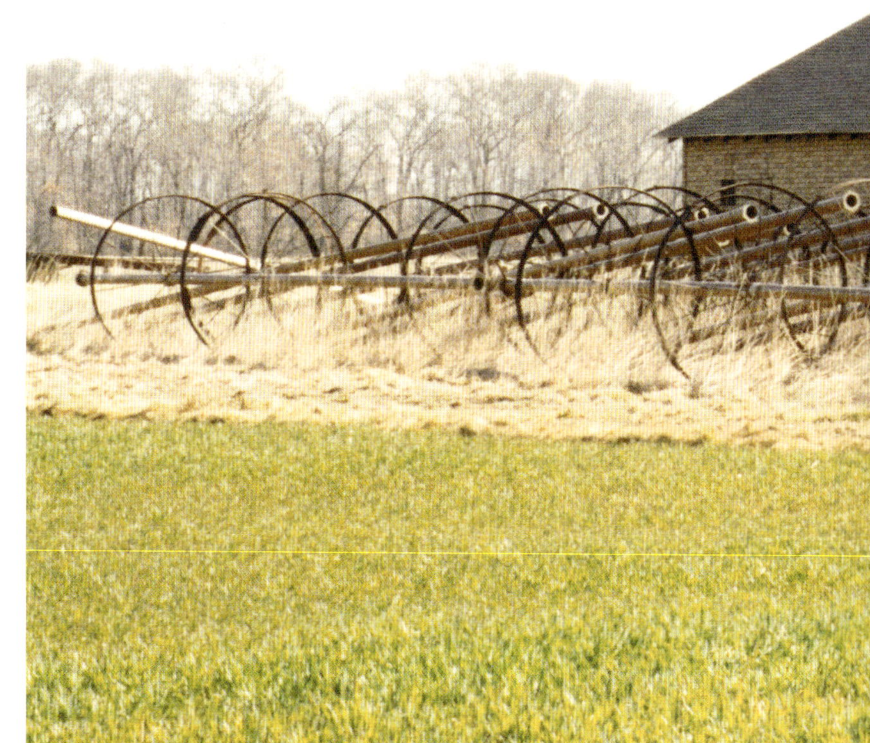

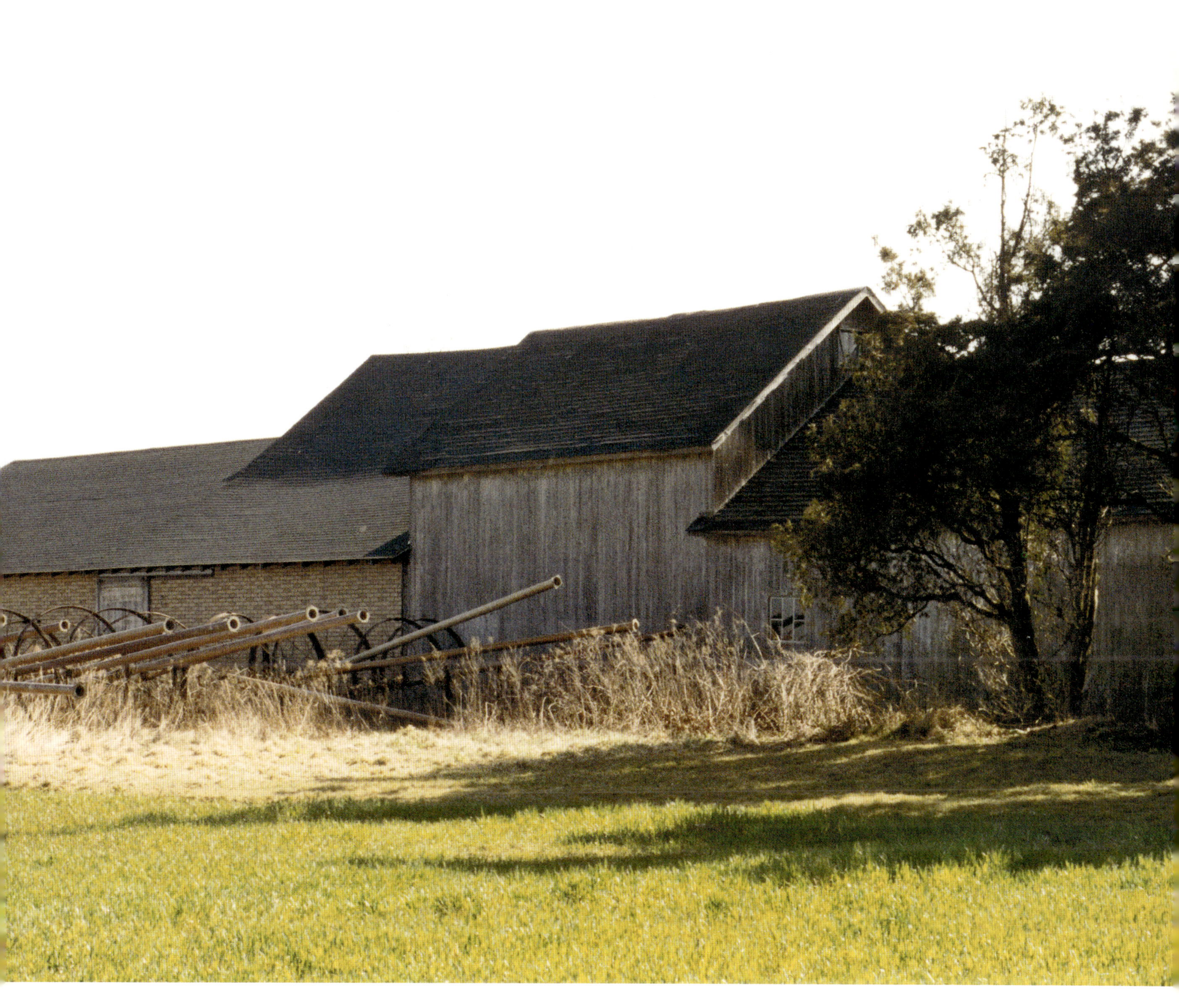

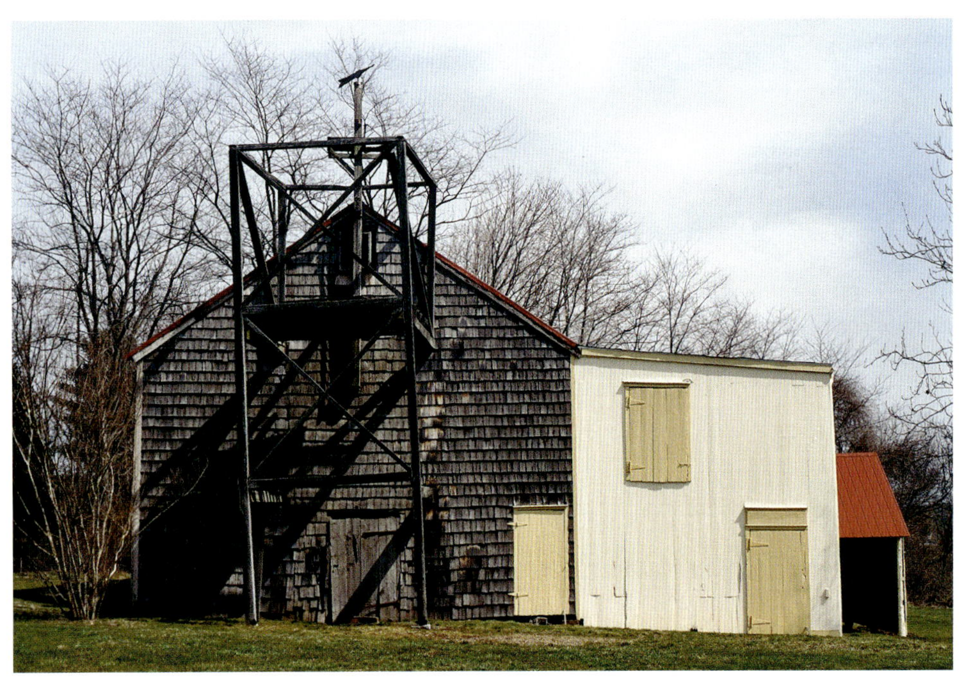

Cutchogue

This unusual barn (above), one of a cluster (right), is believed to have been built in 1852. It contains timbers from the 1732 Cutchogue Presbyterian church that was dismantled in 1850. White oak knee braces flank the center bay, the east wall contains a reused wall girt and a section of a summer beam, and there are notched and pegged roof rafters used in the construction.

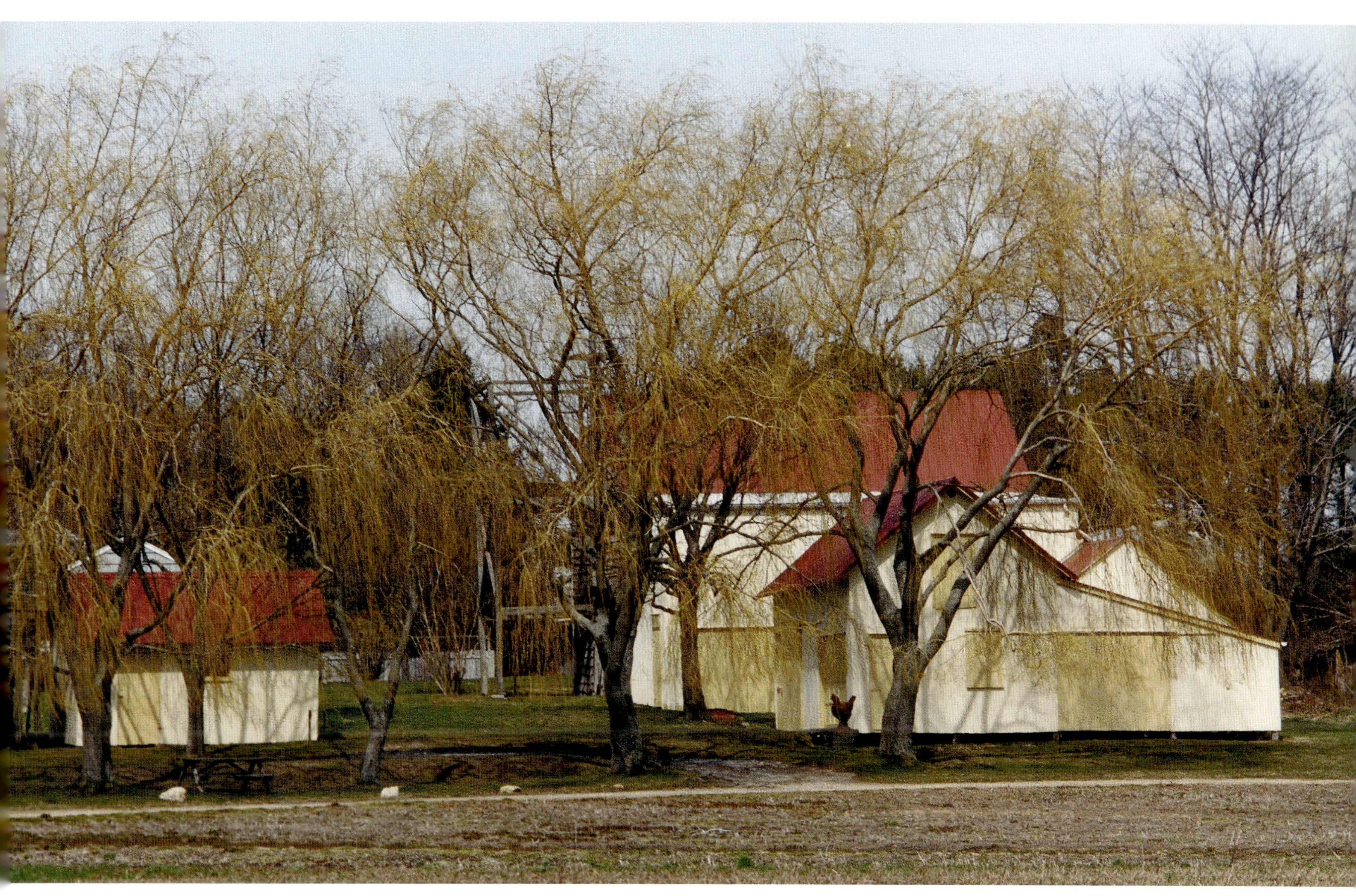

Orient Point
The water tower was added in 1935 to provide for irrigation.

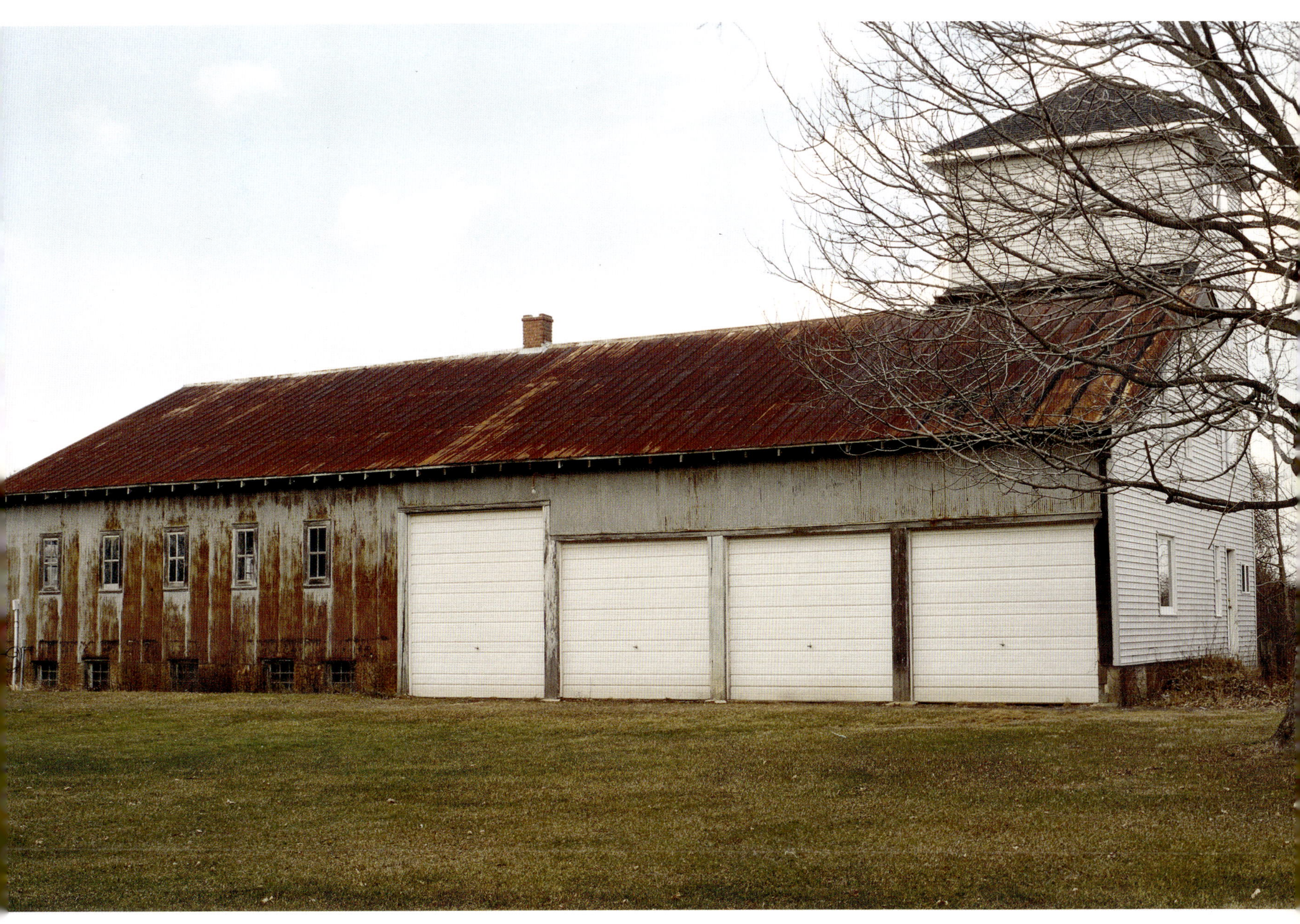

Bayview
This was the carriage house for the
Silas Austin Horton Dayton estate.

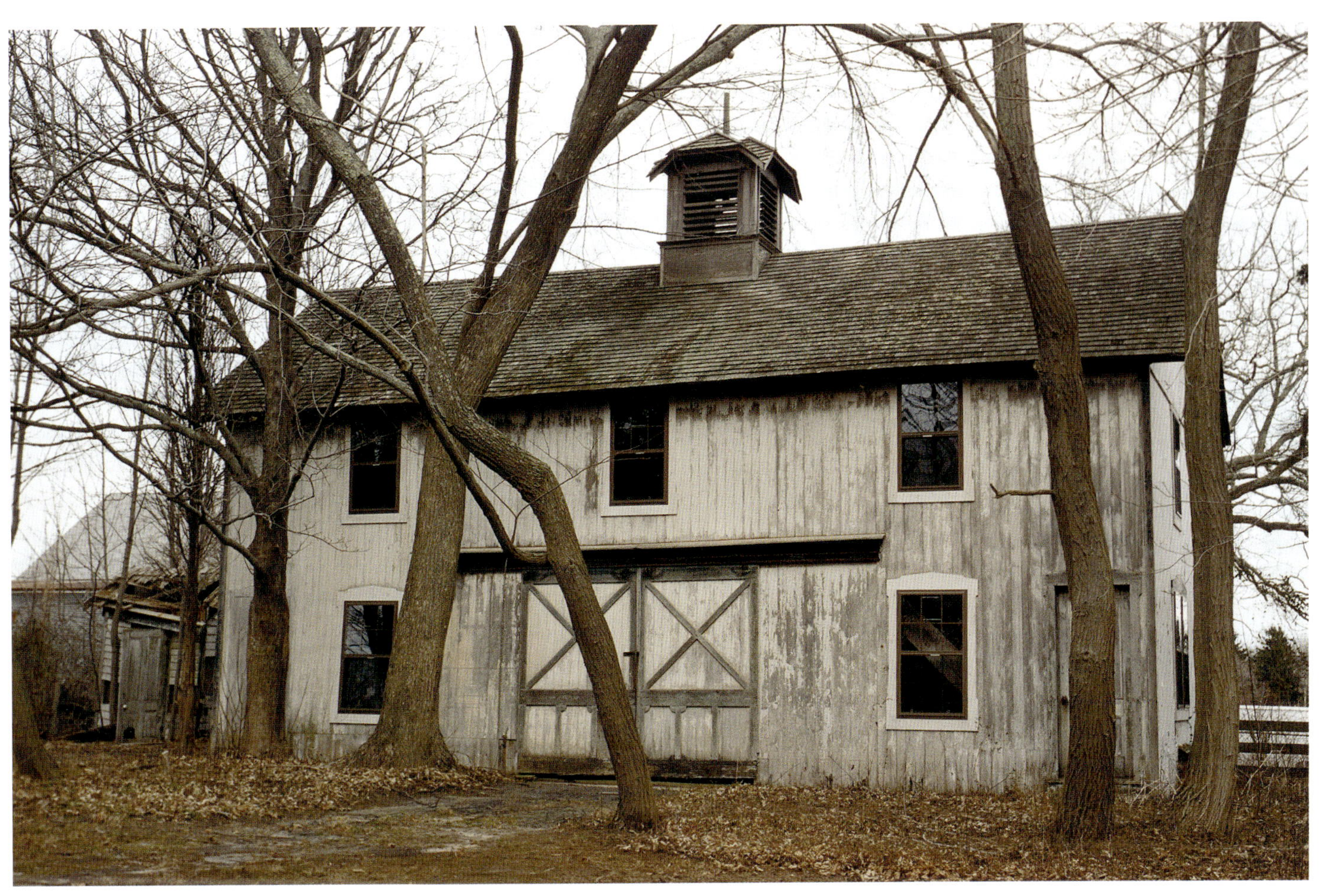

Riverhead

Additions and extensions over time define a working barn.

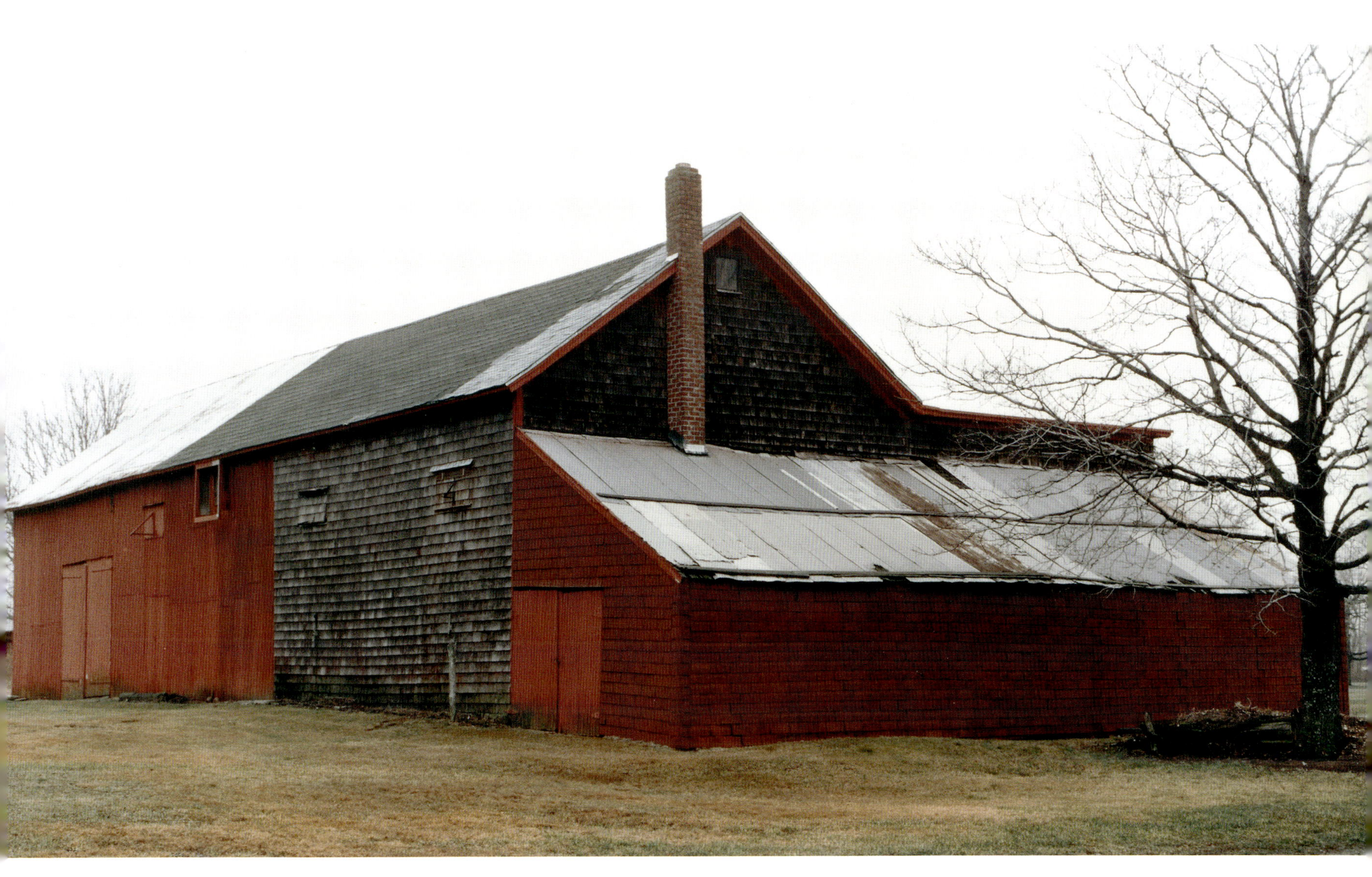

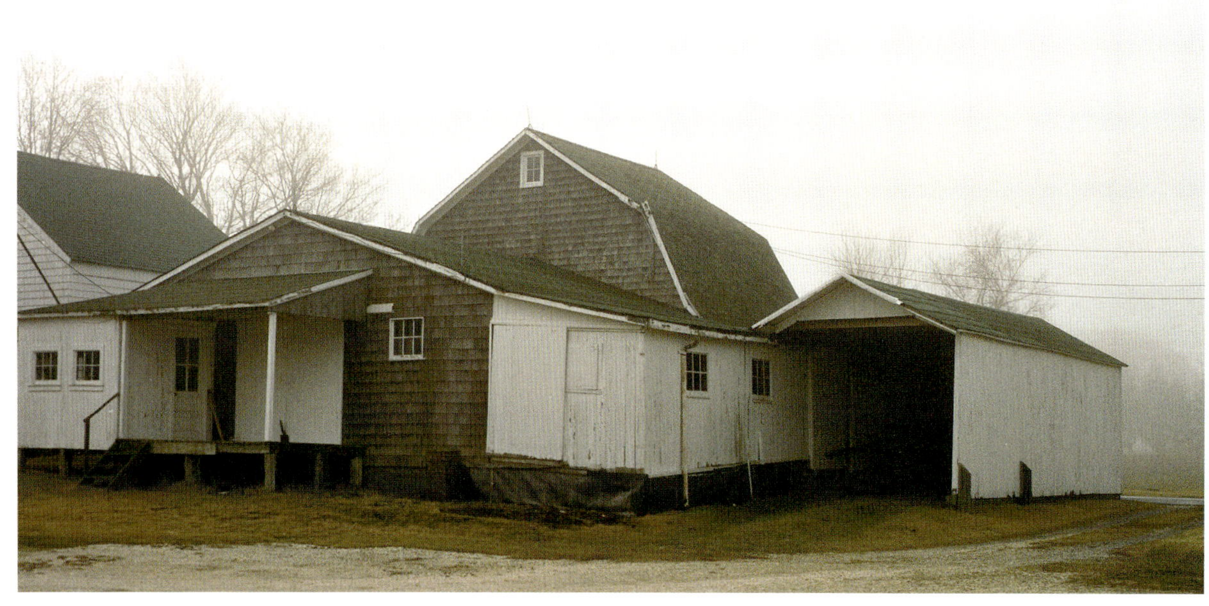

Oregon
The main barn was built before 1900.

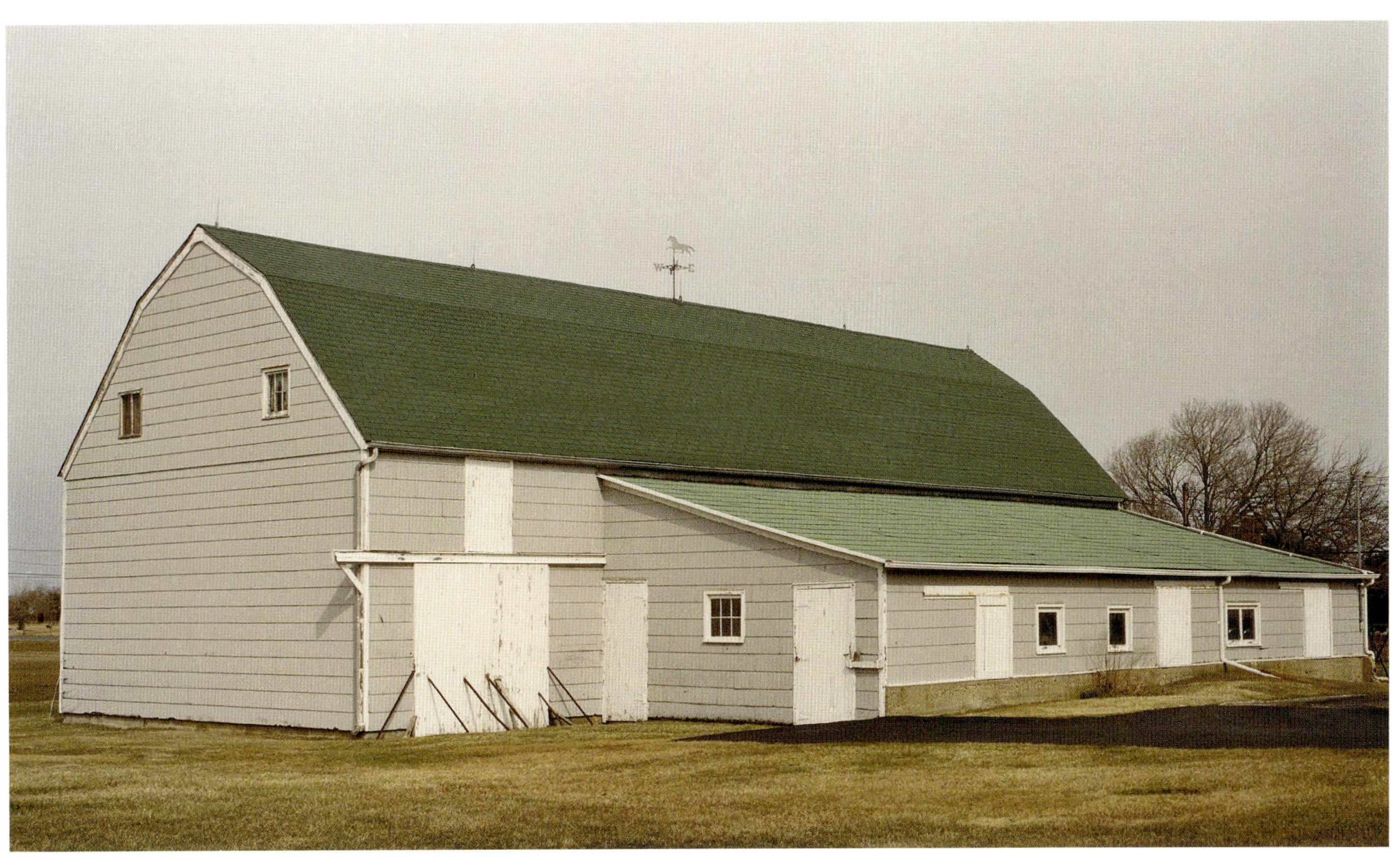

Peconic

It is possible that the barn was built in the early nineteenth century.

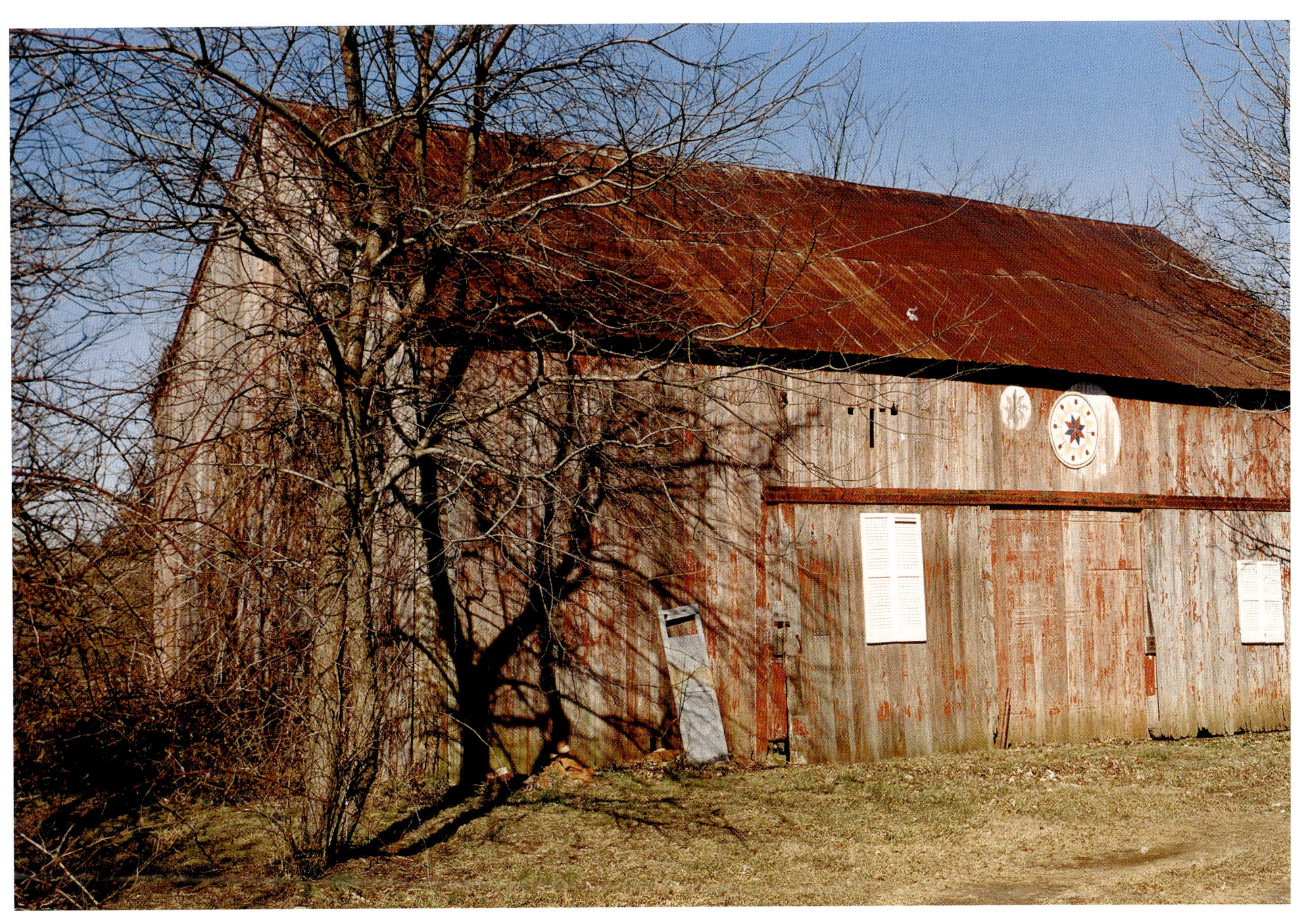

Southold

When a barn's setting changes from rural to residential,
there are few farm outbuildings gracing the barn.

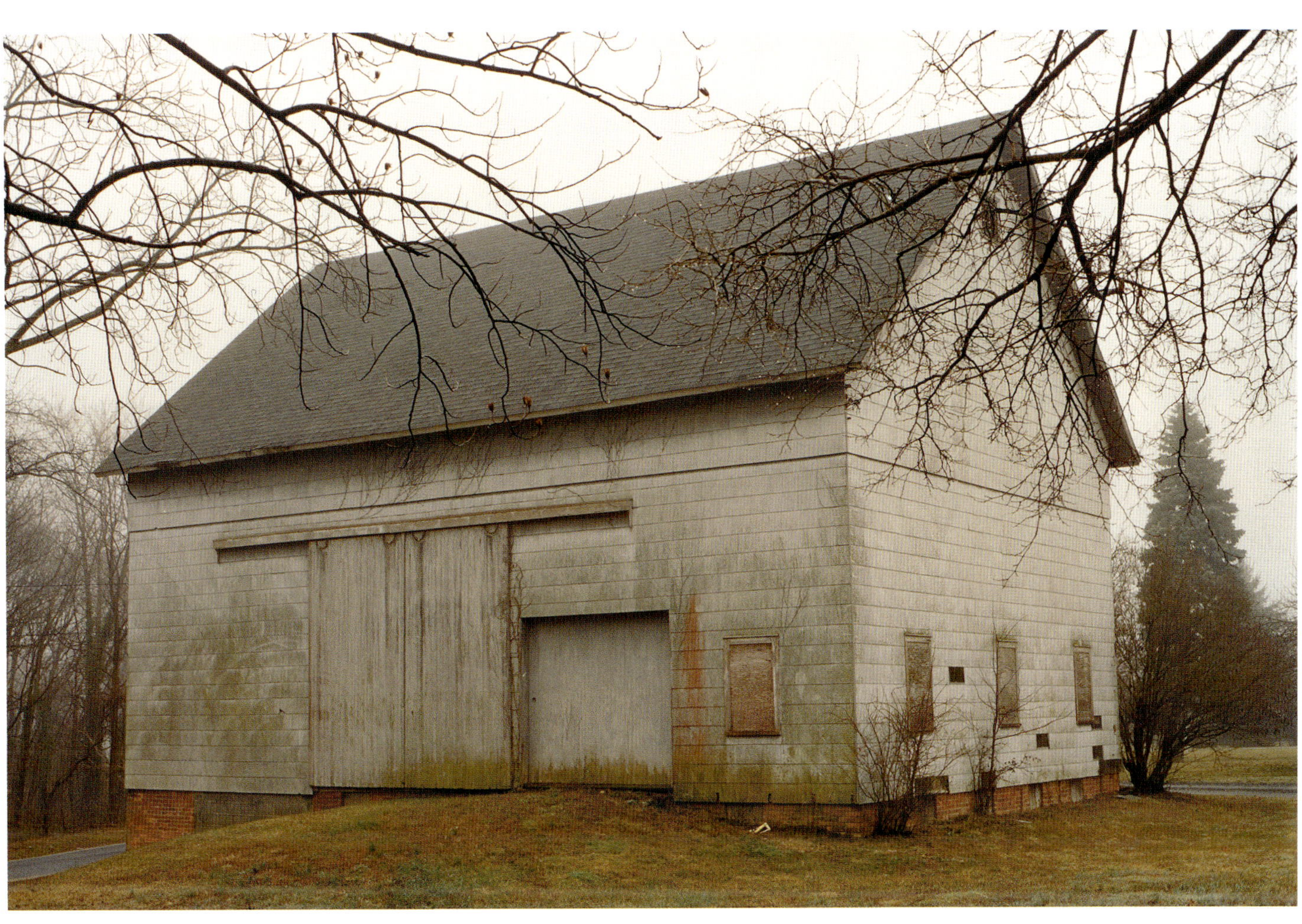

Mattituck
The barn was built in the late nineteenth century.
The farmhouse associated with the barn is gone.

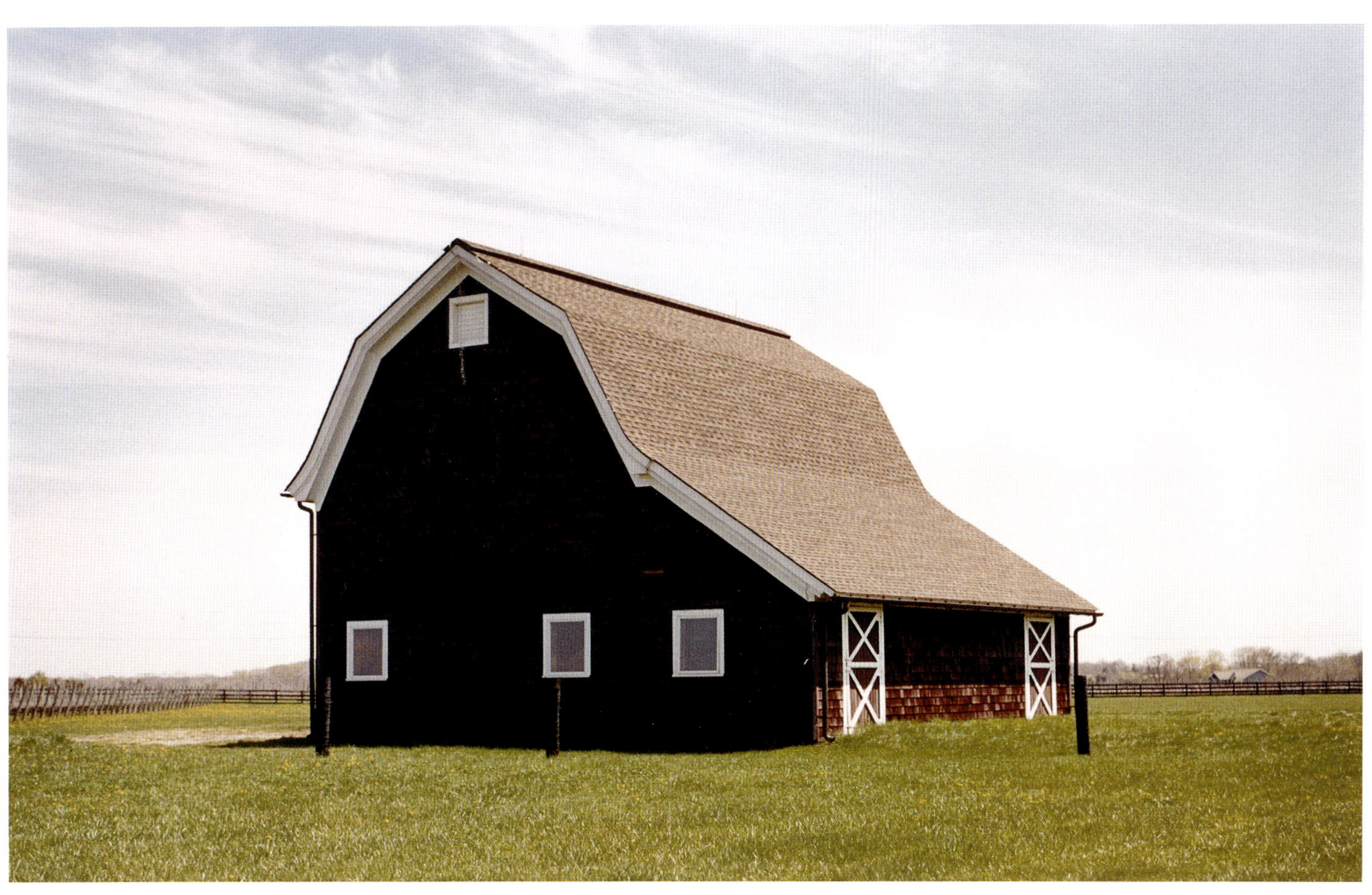

Southold

Barns that remain in agricultural use are modified to accommodate changing needs. The barn on the right was elevated onto a new foundation so that it could store larger farm equipment.

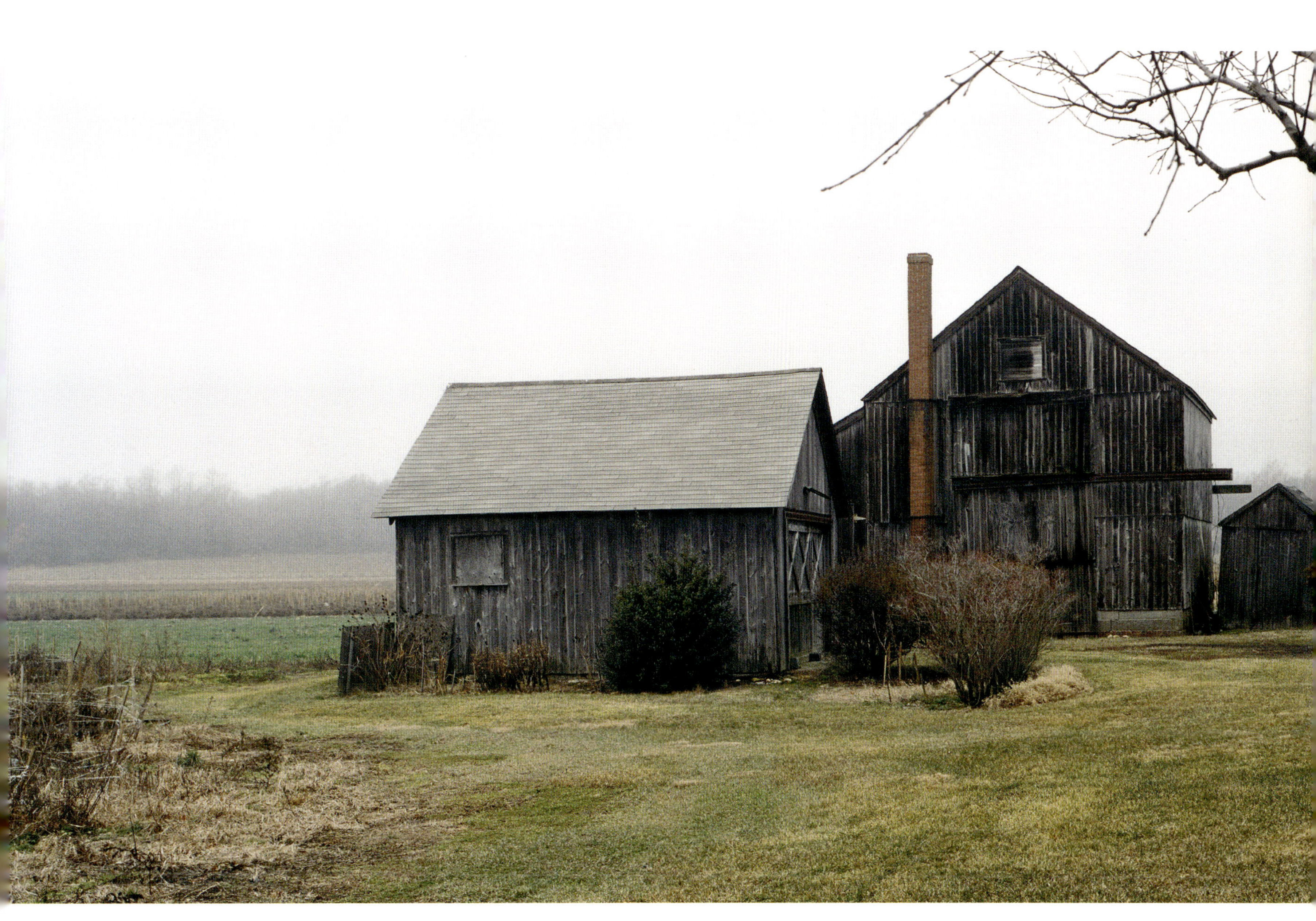

Orient

Carriage houses are usually found in residential settings, which is why they have been called urban barns.

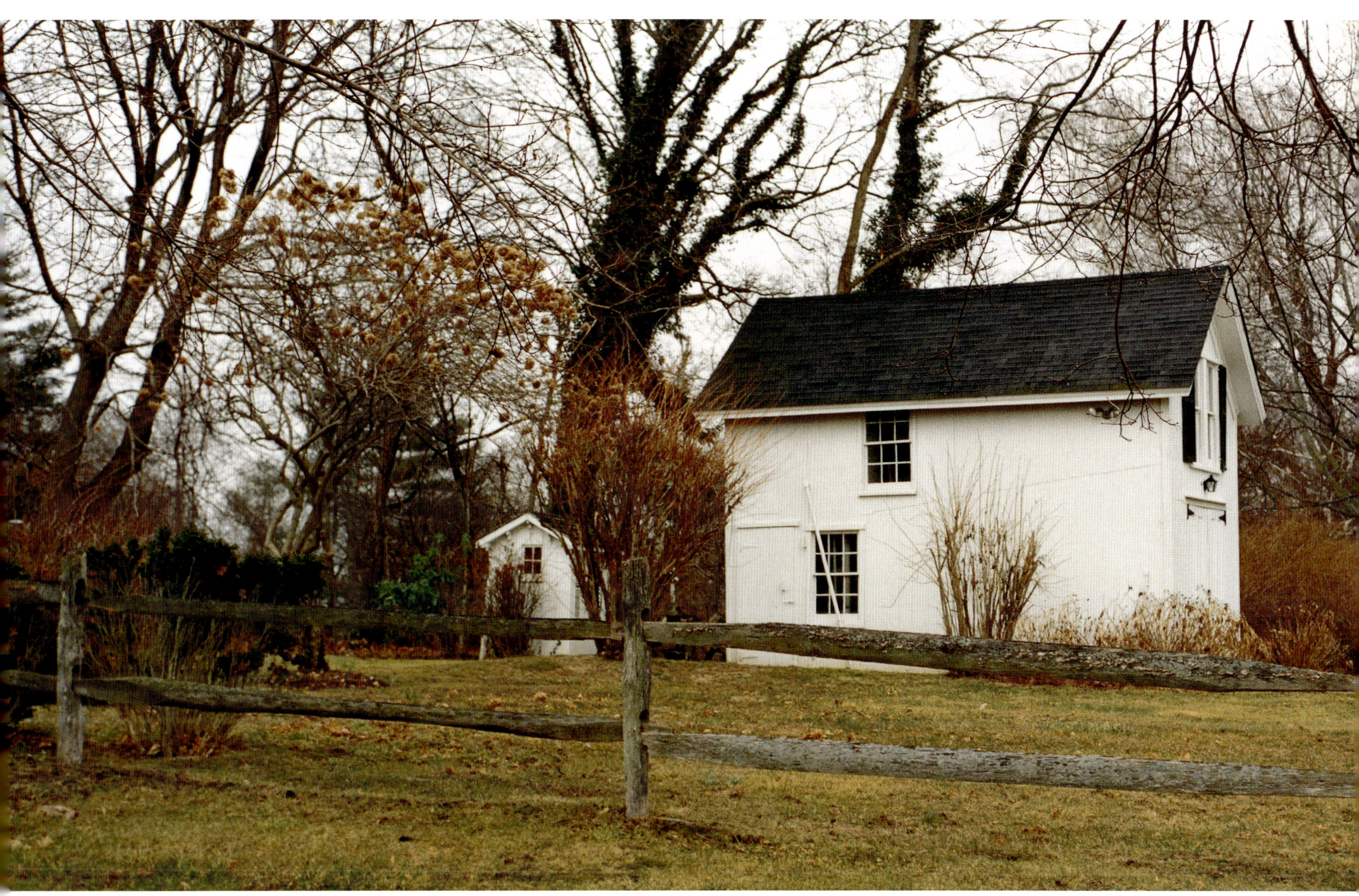

Mattituck

The barn was built in the late nineteenth century.
Two additional vertical plank outbuildings have timbers
from the seventeenth century, including a summer beam.
Site of Mattituck's first windmill (1660–1710).

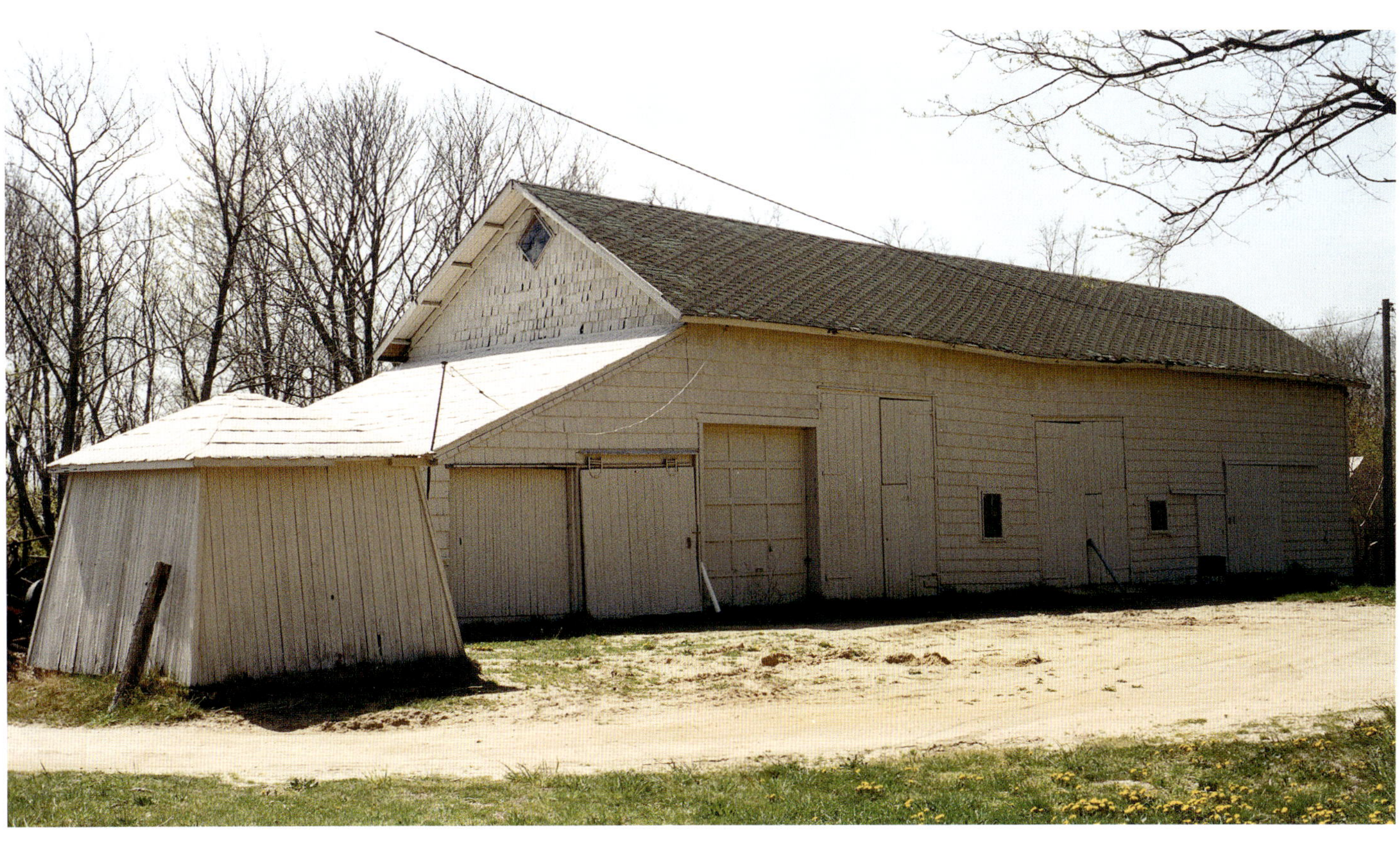

Orient Point

Clapboard is unusual on older barns.
Wood shingles and vertical planks are the preferred sidings.

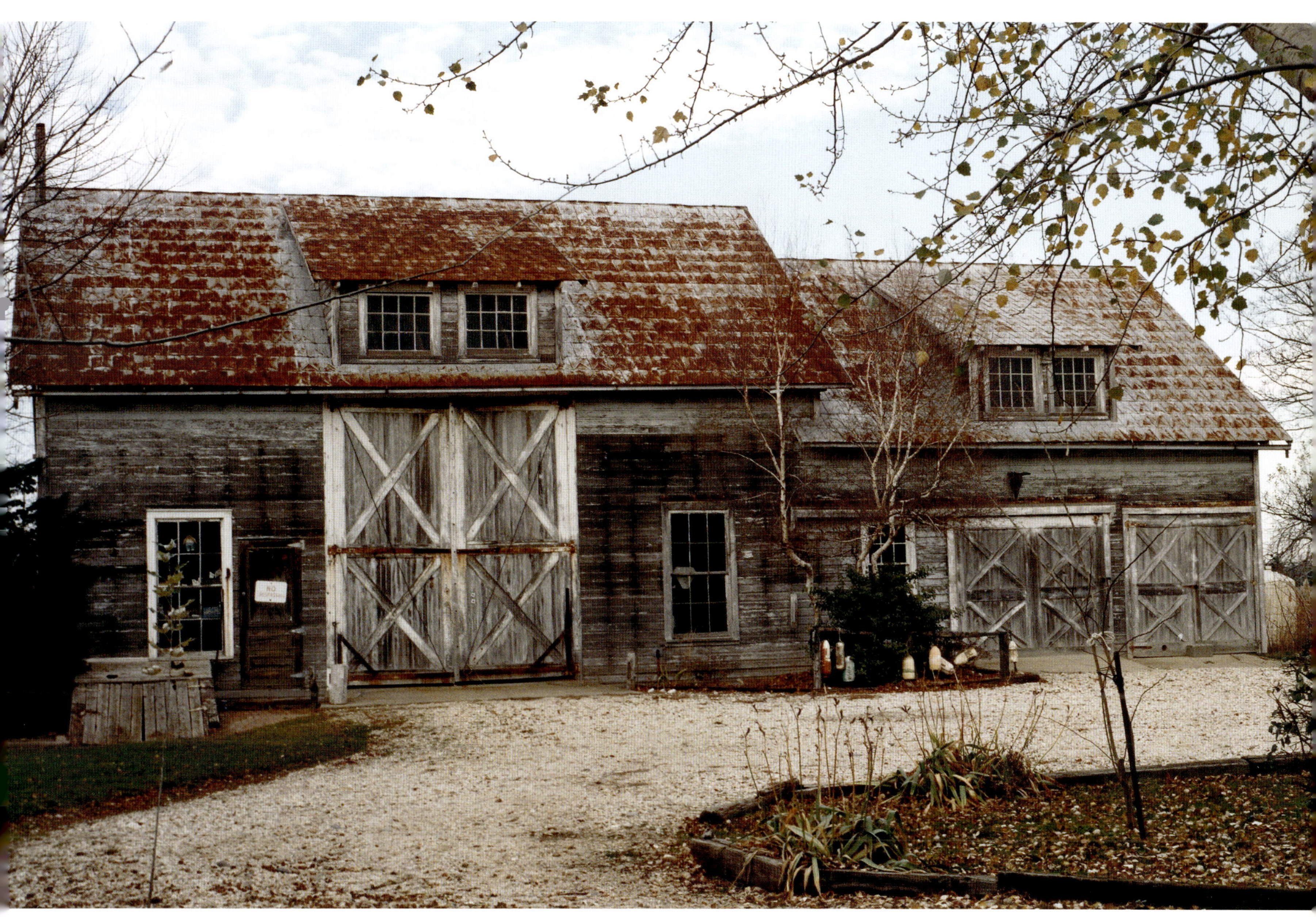

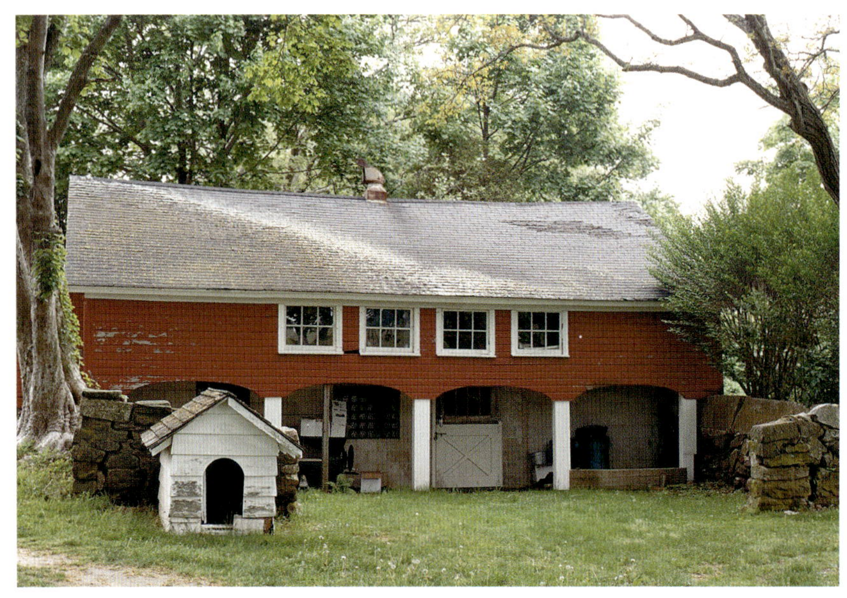

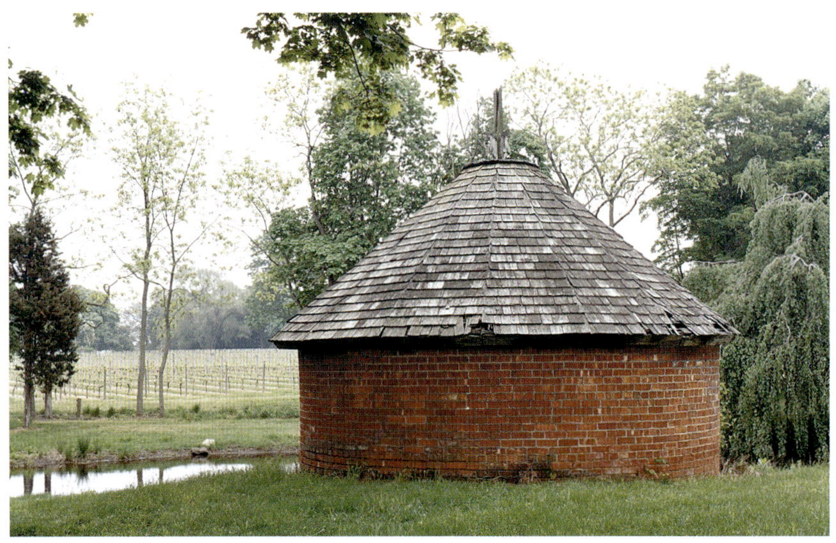

Southold

The main barn (above, top), circa 1860–70, rests on the original stone foundation and was extended for use as a milking barn in 1921.

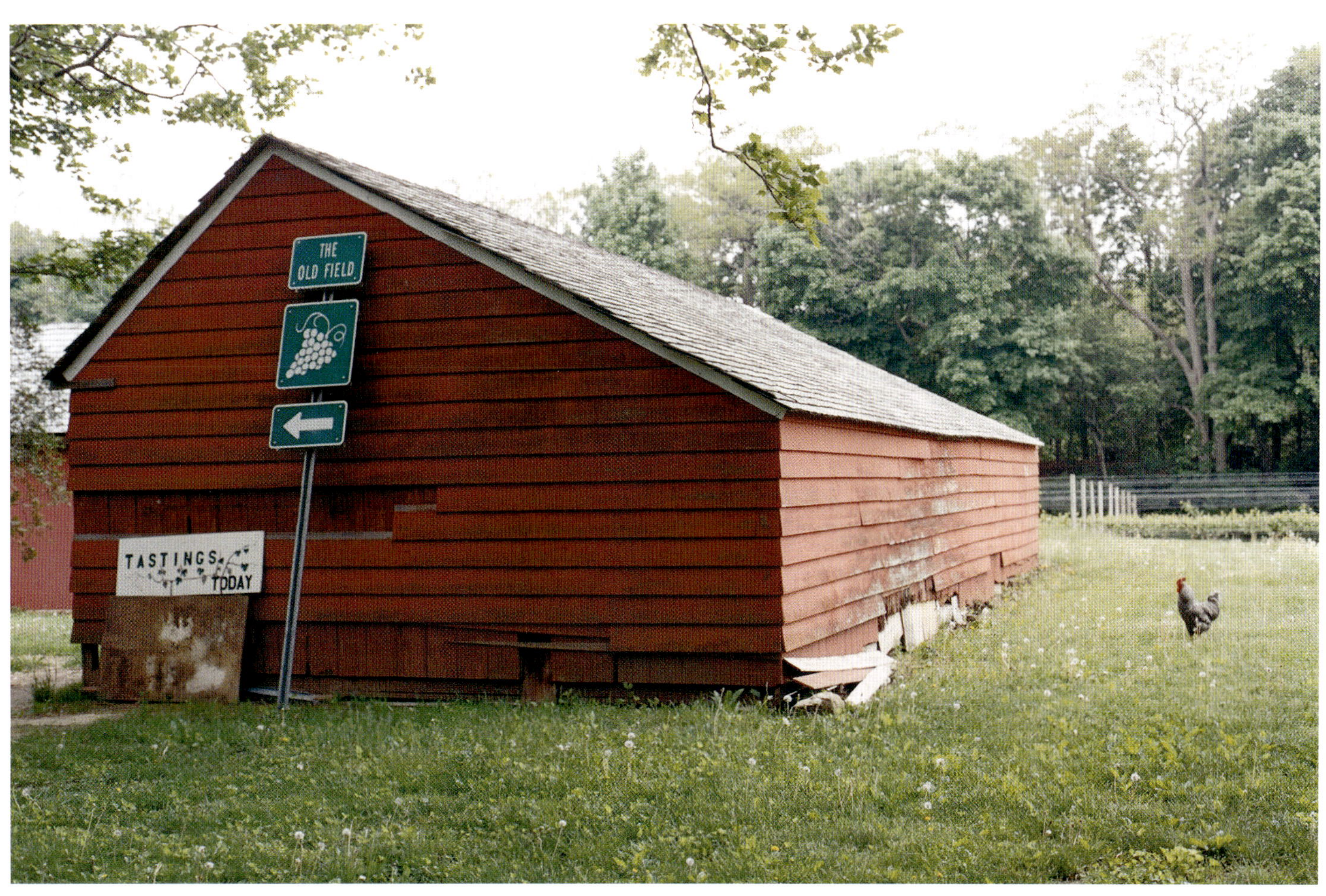
THE
OLD FIELD
TASTINGS
TODAY

Oregon
Buttressing an aboveground barn allows it to handle the weight of a potato crop.

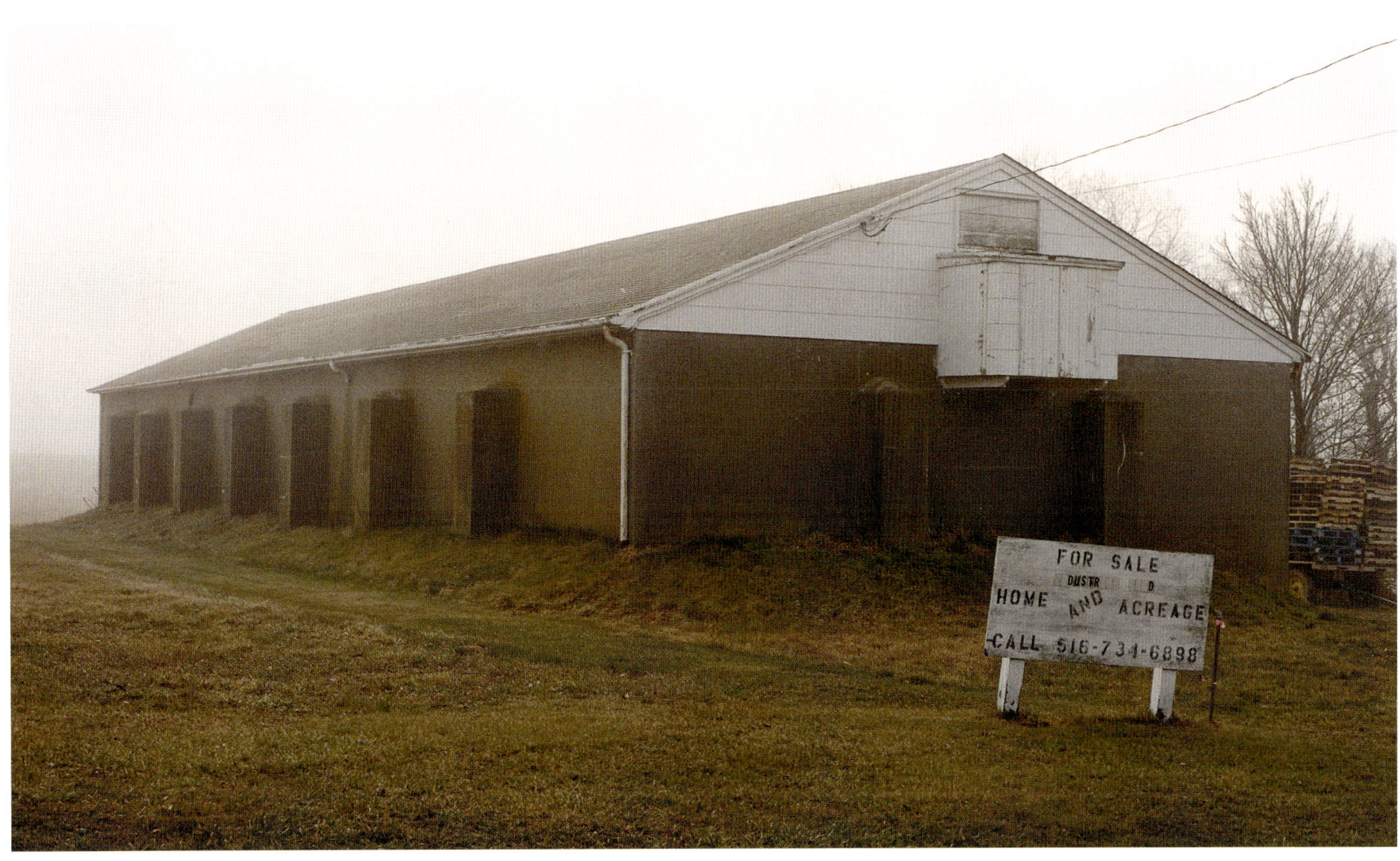
FOR SALE
HOME AND ACREAGE
CALL 516-734-6898

Orient Point
This English hay barn, circa 1900, has a massive frame and a stone foundation.

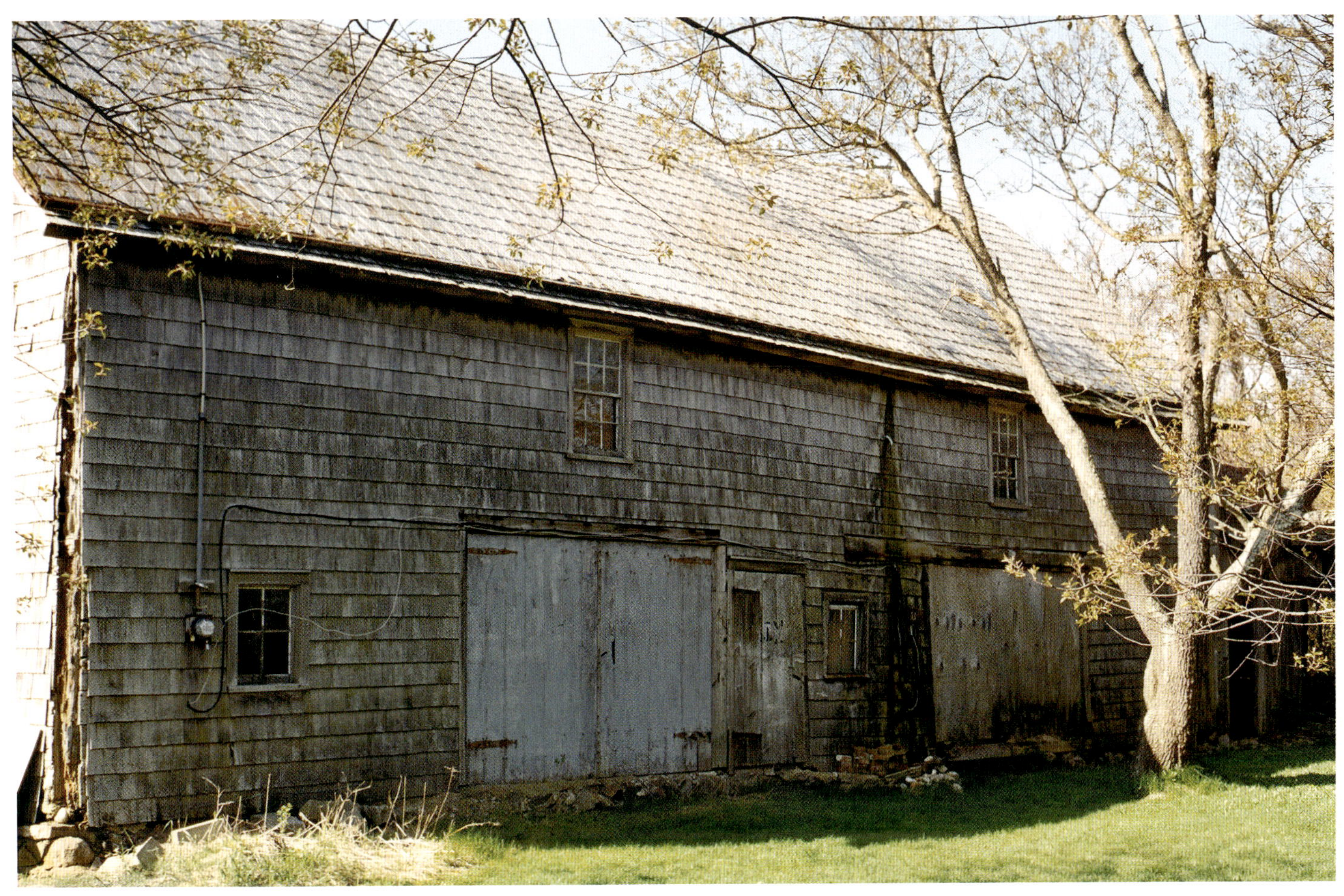

Oregon

Tank houses and gas pumps are attractive reminders of the past.

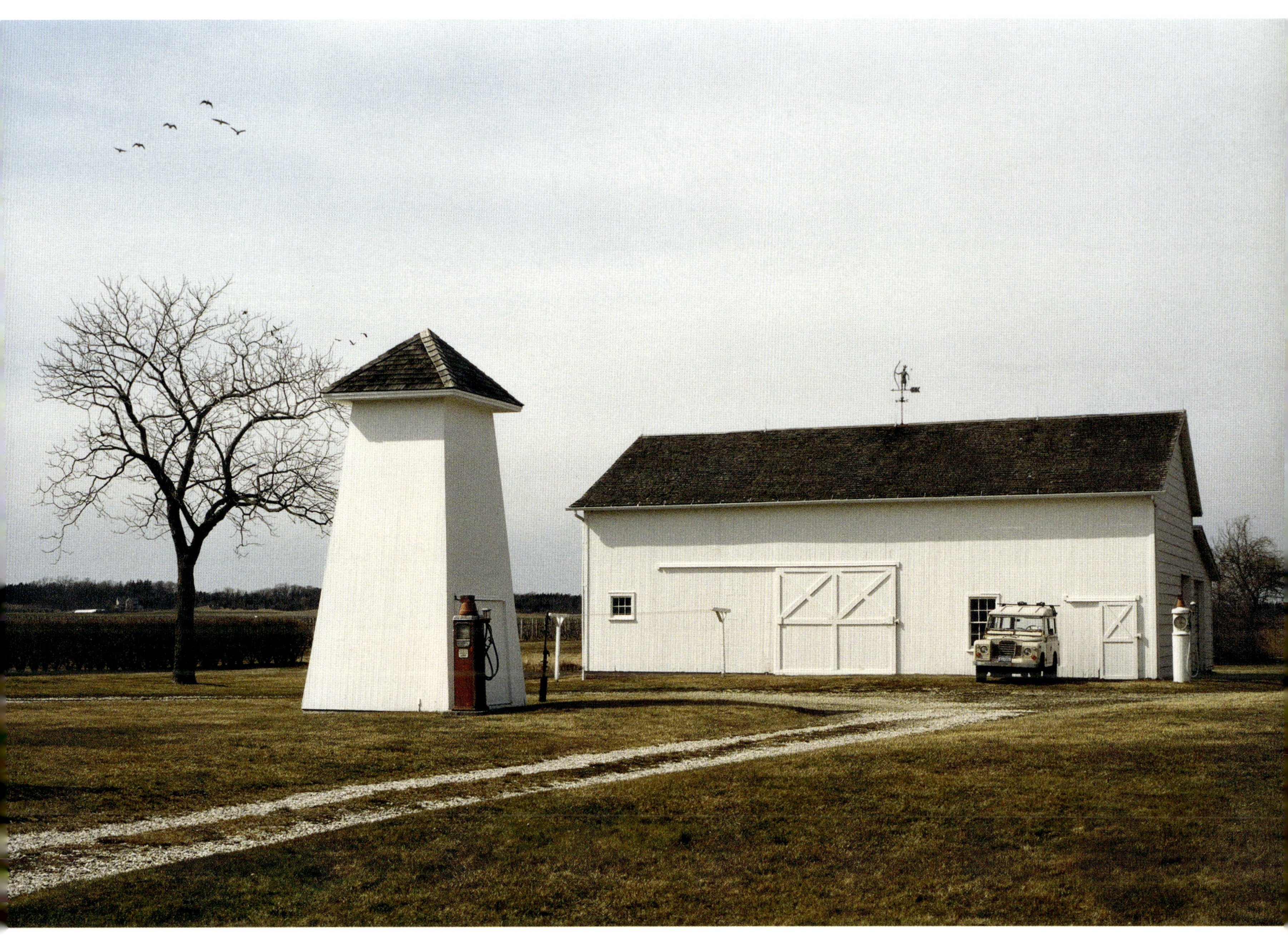

Mattituck

Barns converted to business use are easier to spot than barns converted to dwellings.

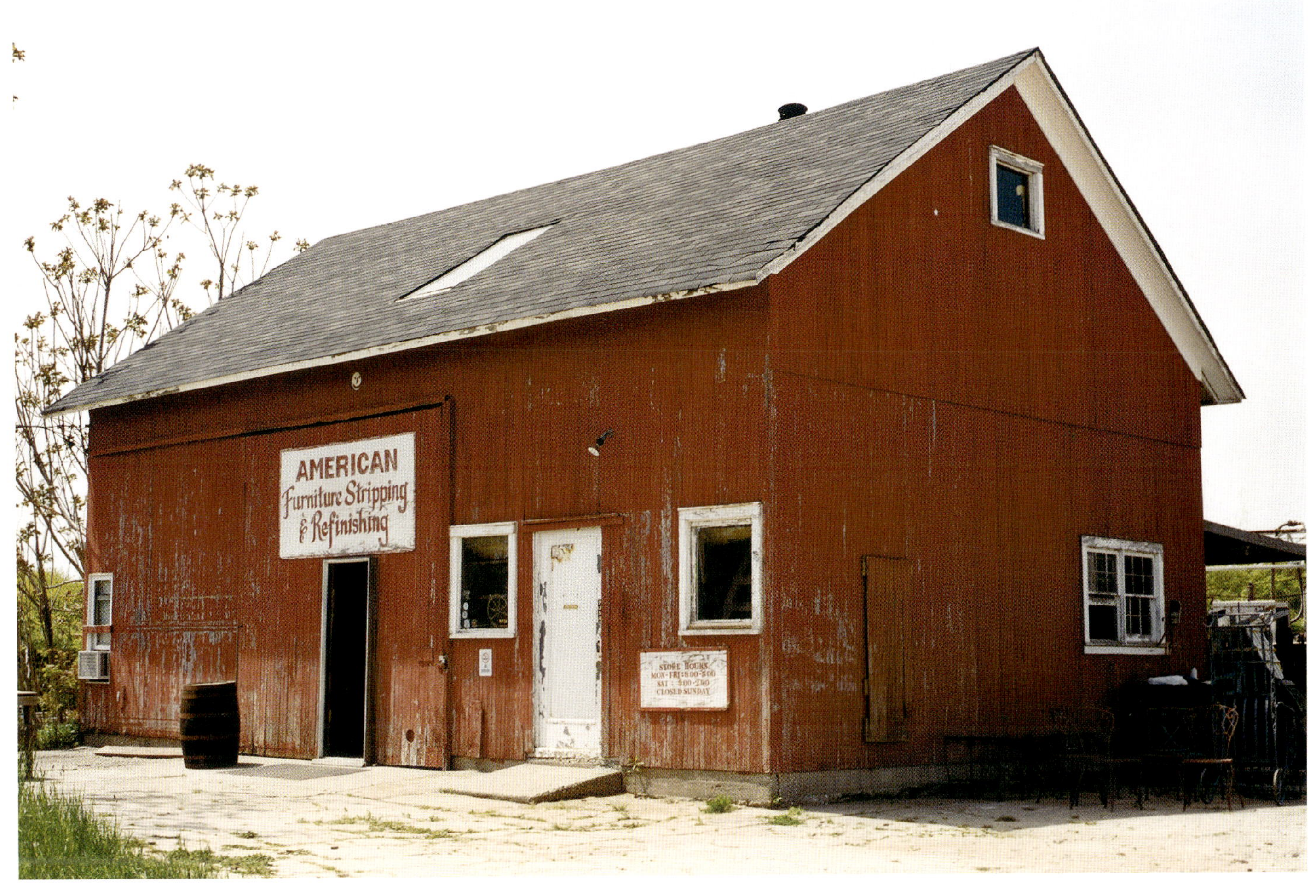
AMERICAN
Furniture Stripping
& Refinishing
STORE HOURS
CLOSED SUNDAY

Riverhead
Sound Avenue is farmland and sometimes the light
on a barn would be too good to pass.

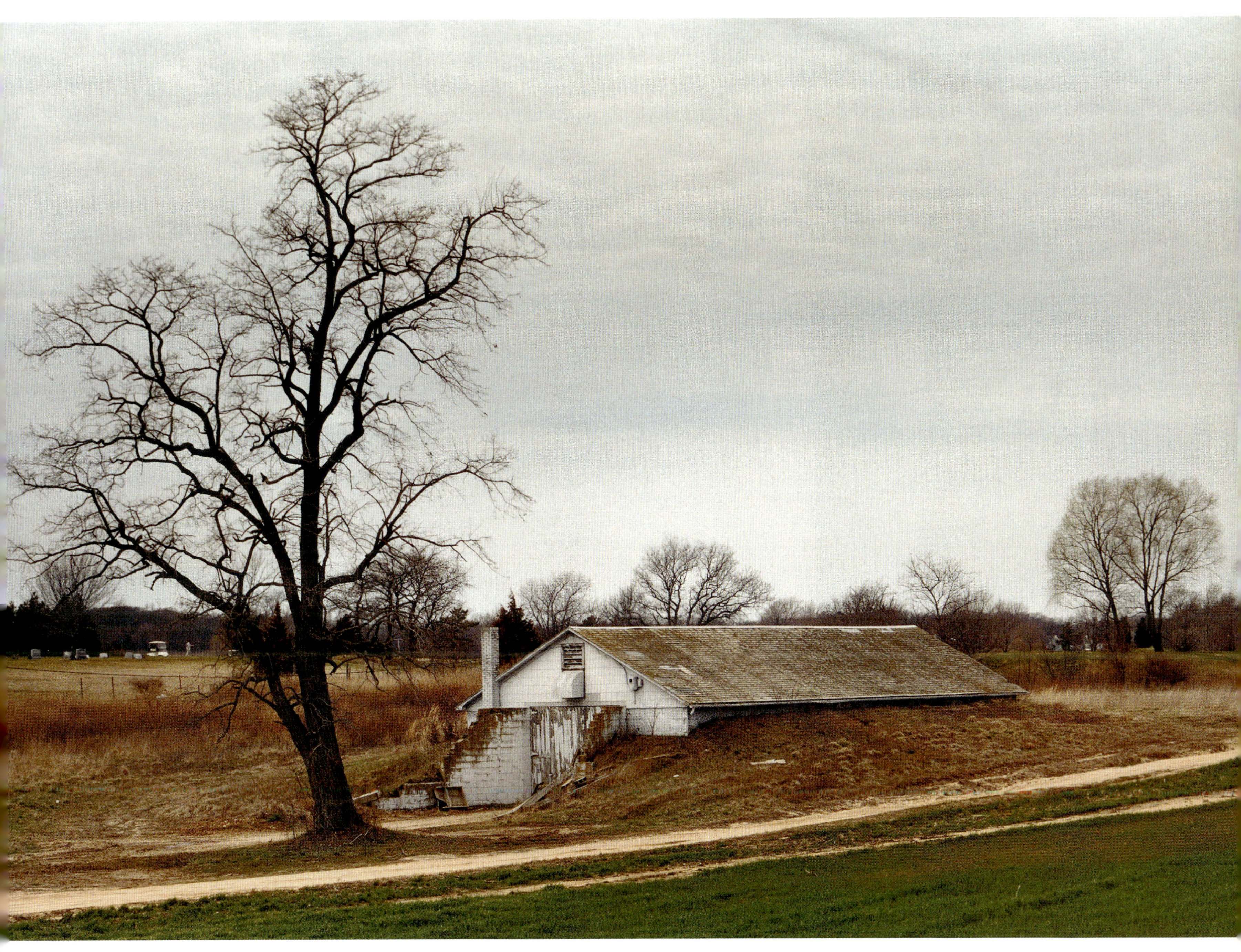

Bayview

The barn and sheds on the property were built in the late nineteenth century.

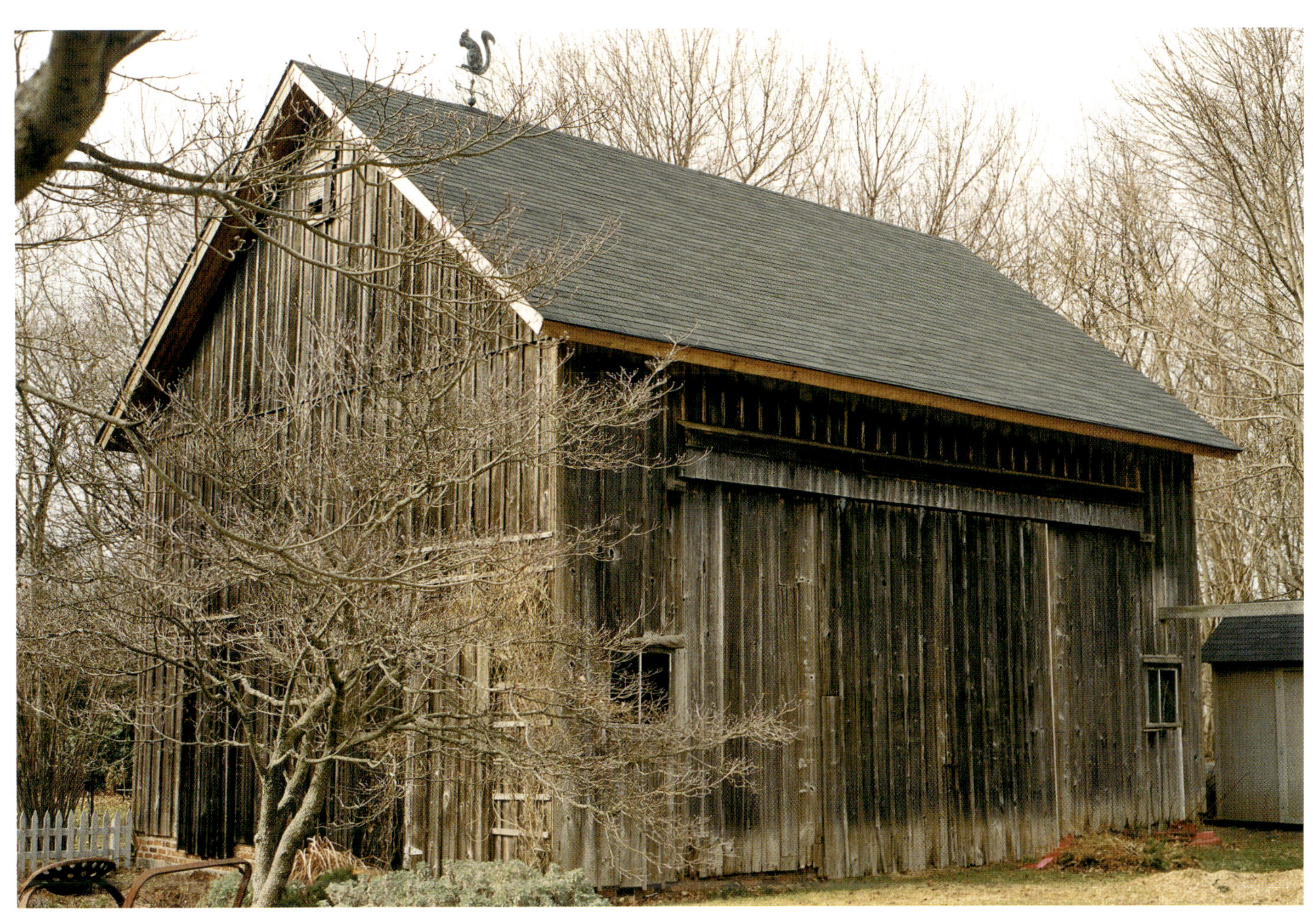

Mattituck

The barn was built in the early nineteenth century. It is pegged and the original wood shingles are still intact under the asphalt-shingled siding.

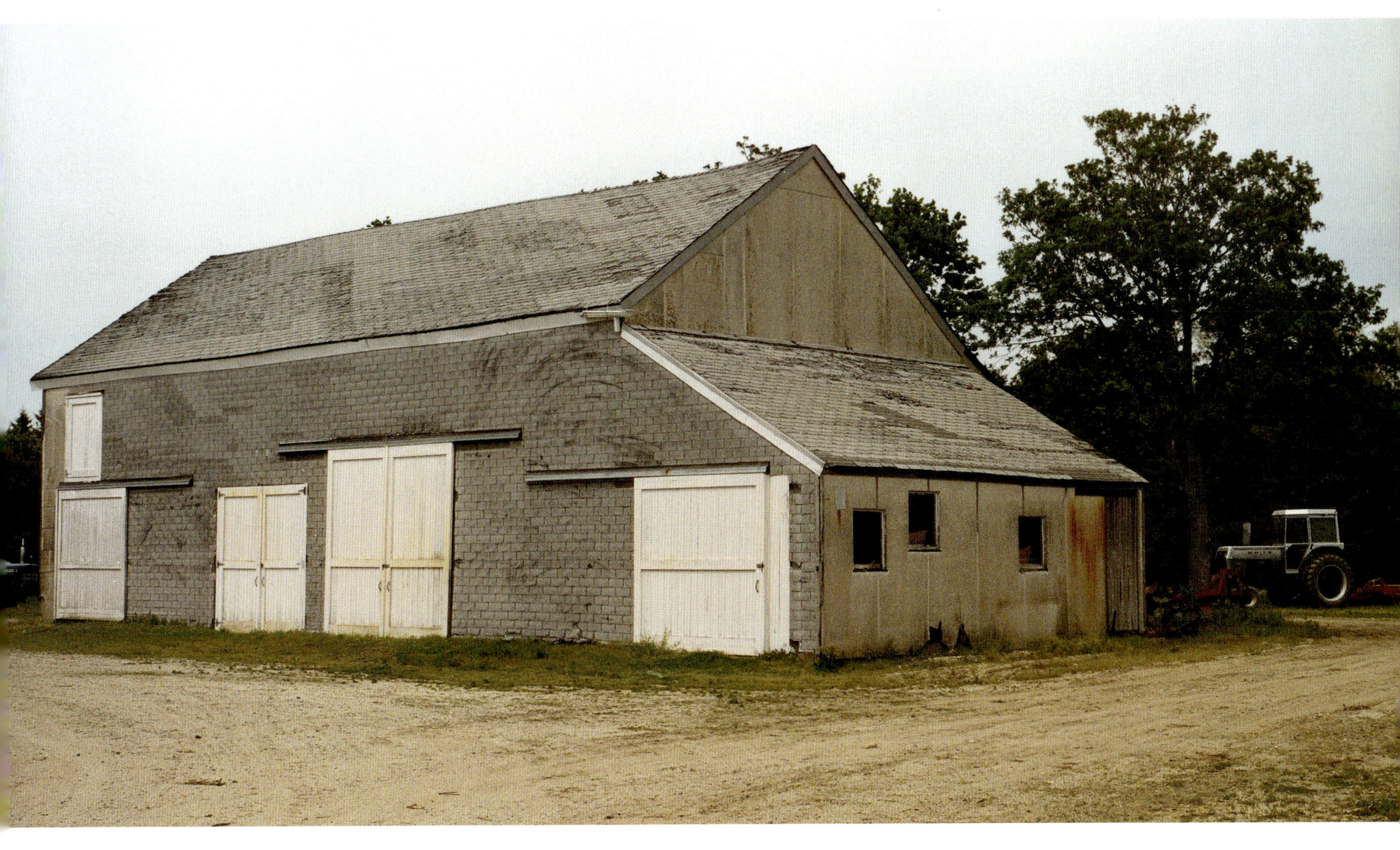

Southold

The original barn burned down in 1918–19 and was rebuilt in 1920. The previous owner, Fran Woodward, painted the sides of the barn as high as she could reach. She was still painting the barn when she was in her eighties. The current owner likes this history and preserves the paint job.

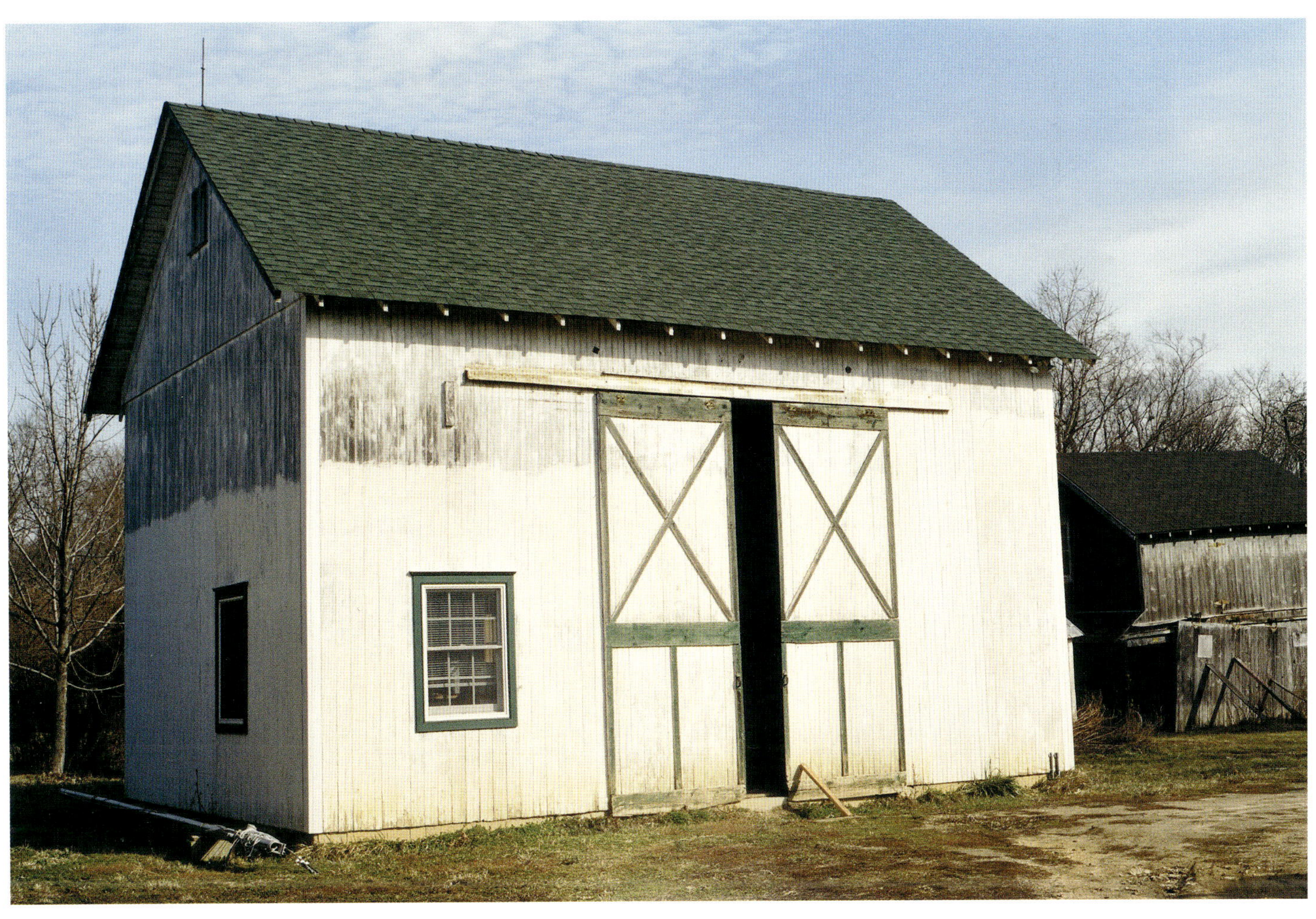

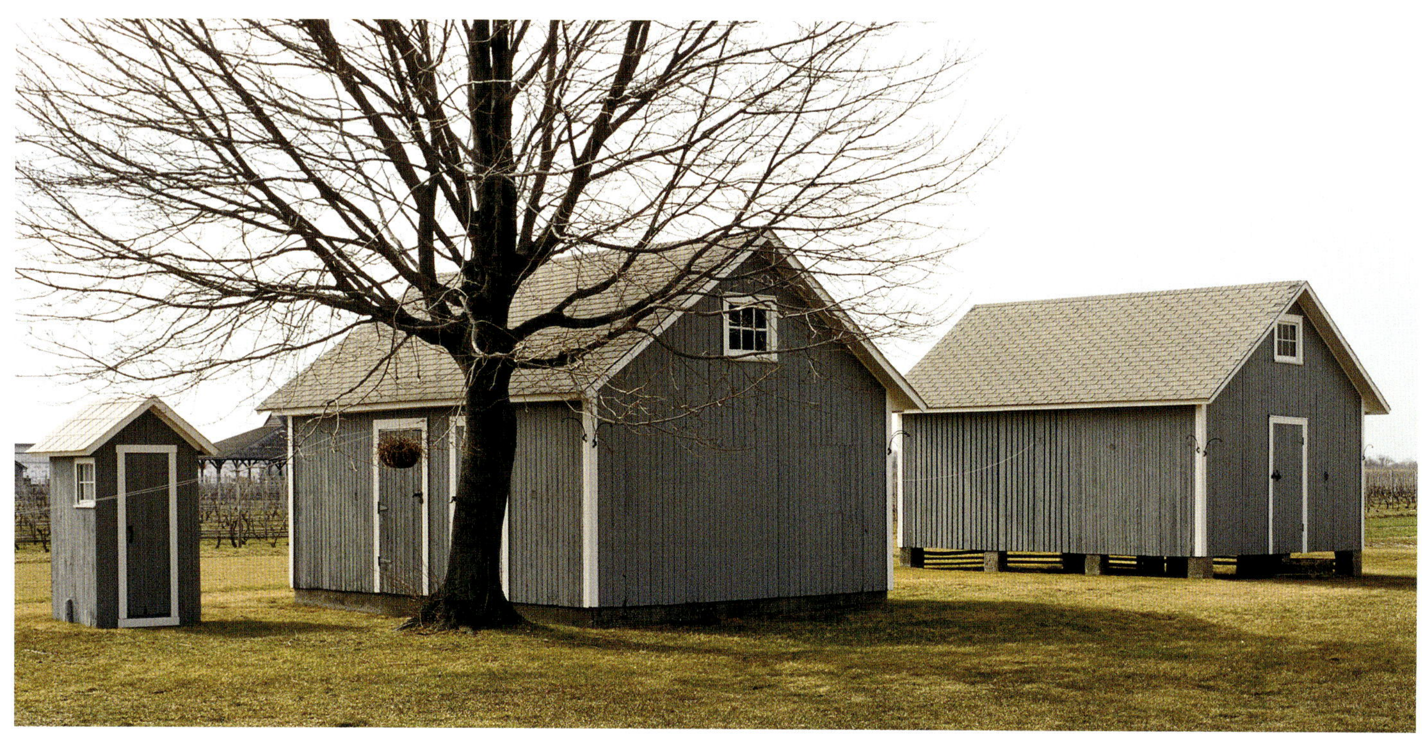

Peconic

This area of Peconic has retained some of its agricultural acreage and many of its wonderful barns. The barn is still in farm use.

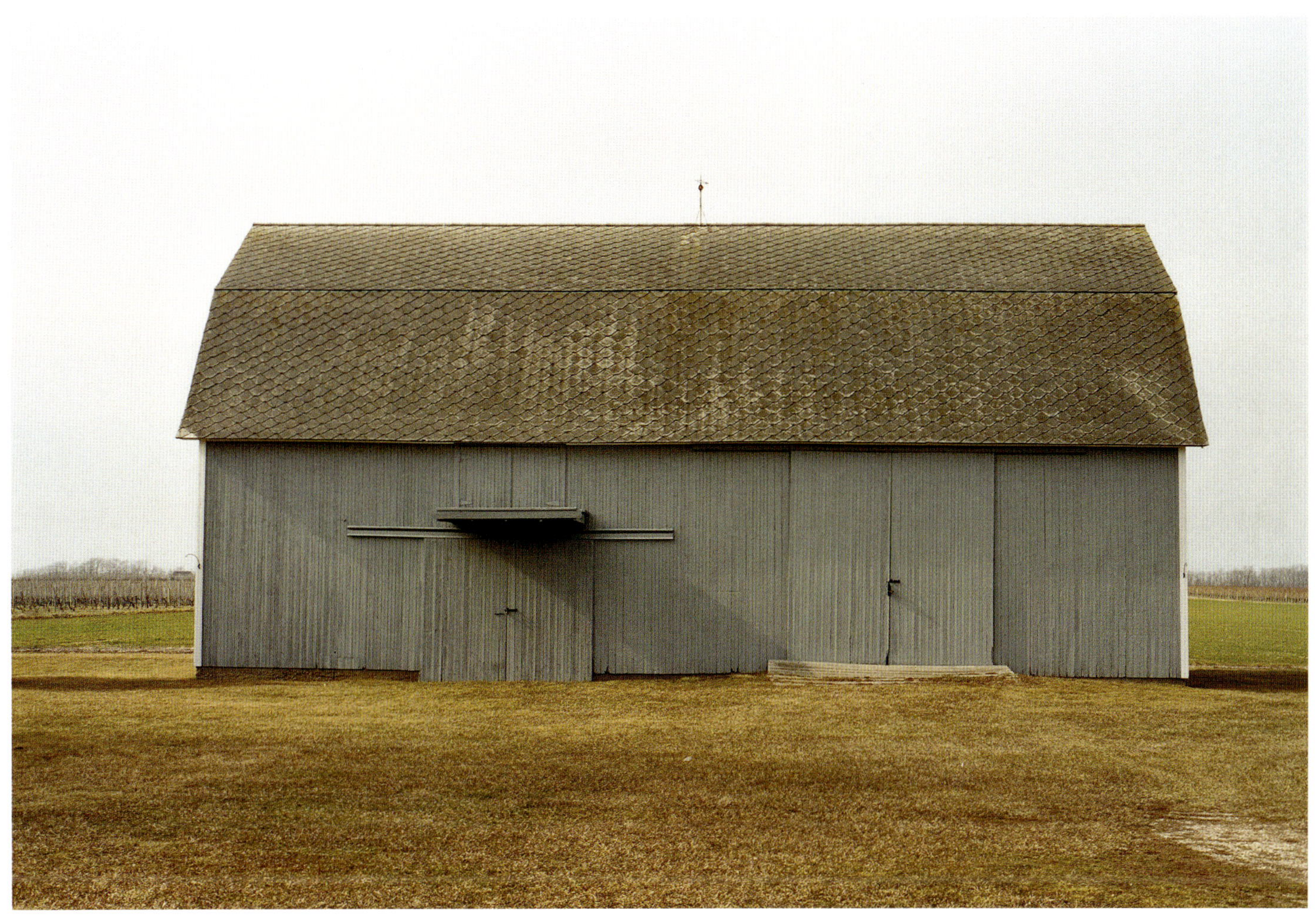

Southold

Barn red remains a popular paint color for barns on the North Fork.

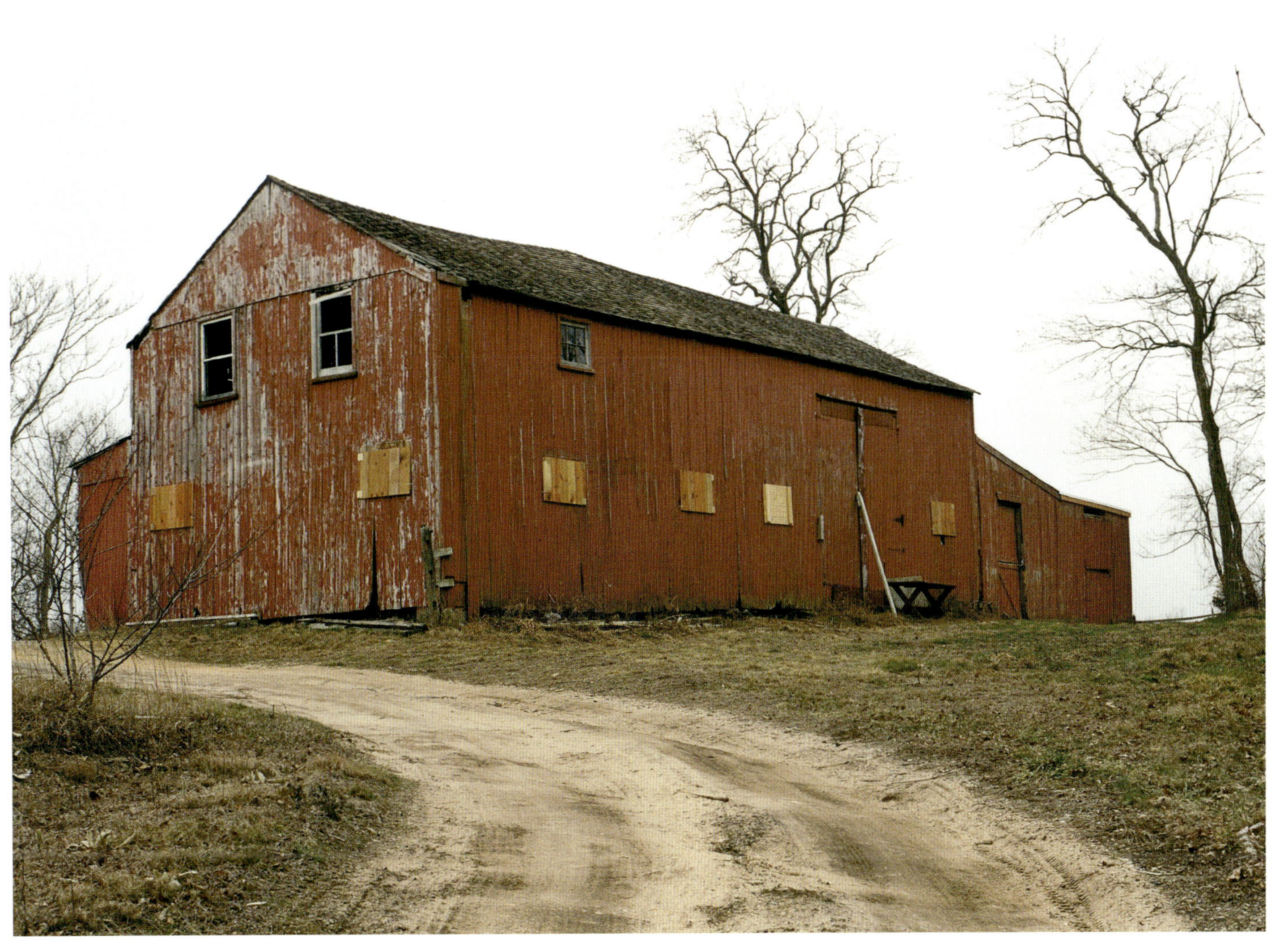

Cutchogue

The picture was taken in 1995 and unfortunately the barn is gone. Barns are disappearing from the rural landscape and new barns are being built. Over time the survey in the town of Southold will allow us to quantify these changes.

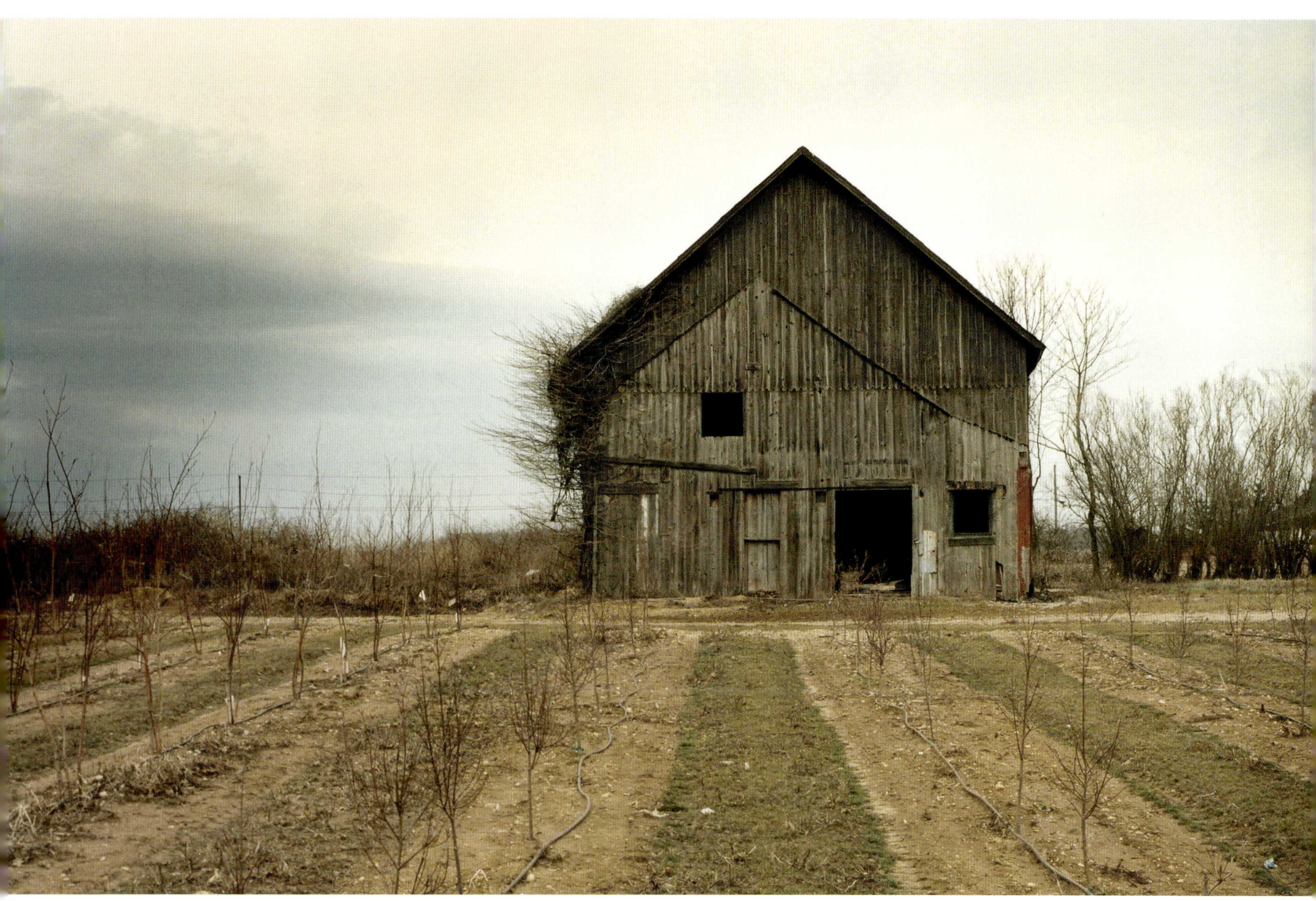

Orient Point, Main Road
Red wood-shingled barn with vertical plank shed addition. 015 9 4.3

Orient Point, Main Road
Wood-shingled early English barn circa 1900, with addition, massive frame, stone foundation. 015 9 7 (page 139)

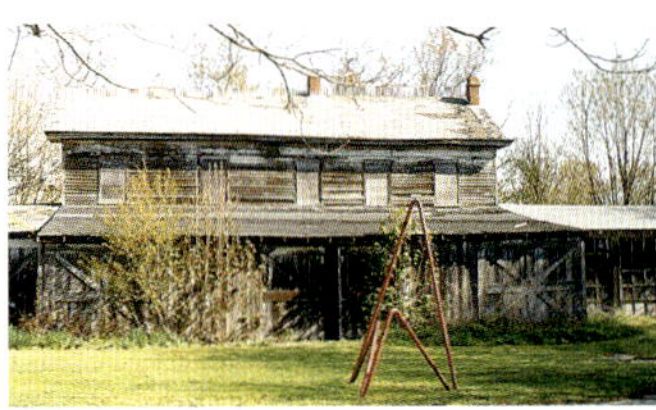

Orient Point, Main Road
Vertical and horizontal plank two-story barn flanked by matching one-story barns. 015 8 14.2

Orient Point, Main Road
Board and batten and wood-shingled barn with shed addition. 015 8 1.1

Orient Point, Cedar Birch Road
Vertical plank saltbox-shaped barn with addition. 015 8 26.8

Orient Point, Main Road
Red horizontal plank barn, with white shed addition. 020 3 9.3

Orient Point, Main Road
Small red vertical plank garage or carriage house circa 1900. Large English barn gone from site. 015 2 17.4

Orient Point, Main Road
Green vertical plank barn with wood-shingled shed addition; green vertical plank and wood-shingled double barn with addition; white plank barn with additions. 015 2 15.6, 020 3 7.1

Orient Point, Main Road
Vertical plank double barn; smaller sectioned outbuilding; vertical plank chicken coop. The barn was moved from Terrywald, the seventeenth-century house next door, before the 1820s. It is a three-bay English hay barn with eighteenth-century framing. The center bay wide board floor has been preserved. 020 1 3.2 (page 39)

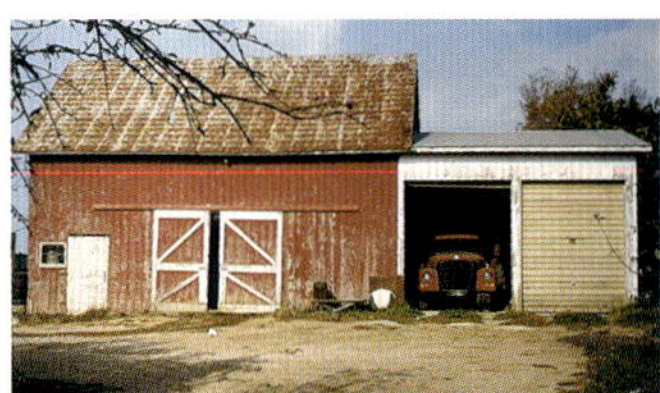

Orient Point, Main Road
Red vertical plank English barn (nineteenth century) with wrap-around shed addition (1970s). 020 1 1.3

Orient Point, Main Road
White vertical plank barn with addition. 019 2 9.1

Orient Point, Main Road
Gray horizontal plank double barn with red tin roof. 019 2 8.1 (page 133)

Orient Point, Main Road
Gray corrugated-tin saltbox barn with shed addition; gray stucco bookend barns. 019 2 7.1 (page113)

Orient Point, Main Road
Green asphalt-shingled barn; two smaller green asphalt-shingled barns. 019 2 4.13

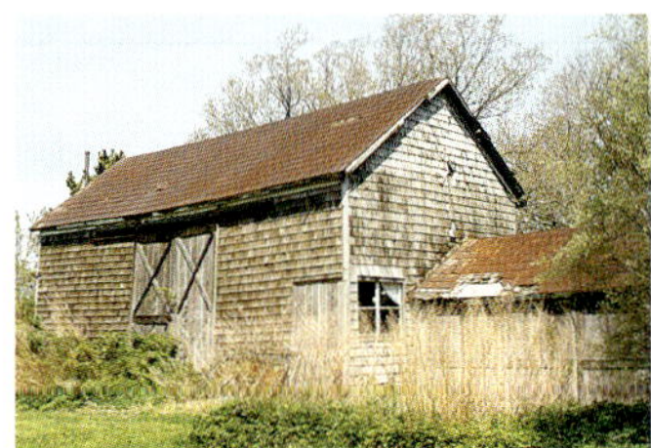

Orient Point, Main Road
Wood-shingled and vertical plank double barn. The three-bay hay barn portion of the double barn was moved onto its stone cellar. Small vertical plank barn. 019 1 14.5

Orient Point, Main Road
Vertical plank double barn with shed addition. 019 1 14.4

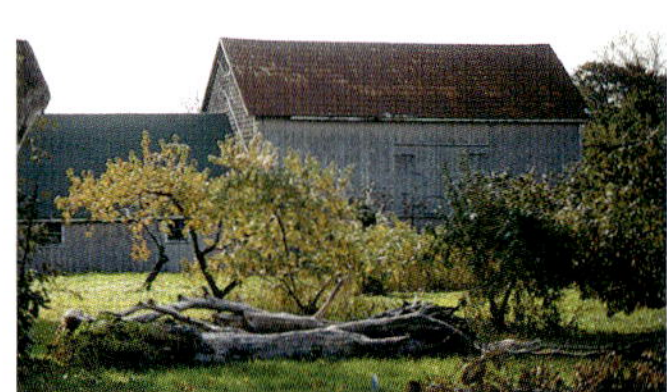

Orient Point, Main Road
Vertical plank double barn. 019 1 11.3

Orient Point, Main Road
Red vertical plank barn; collapsing barn; green vertical plank barn. 019 1 10.5 (page 51)

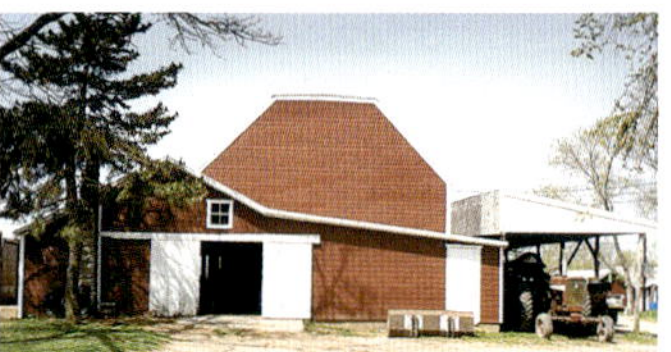

Orient Point, Main Road
Red asphalt-shingled double English barn circa 1840 with shed addition; vertical plank barn late nineteenth century; two gray square tin barns; Quonset hut barn. The house was built in 1849, which lends credence to the date of the English barn. 019 1 8.4

Orient Point, Old Main Road
Farm building complex. The large barn is gone, but various farm buildings remain: chicken coop (1927) and cauliflower shed and sprout house, both built before 1909. 014 2 3.4

Orient Point, Main Road
White asphalt-shingled and vertical plank barn; white wood- and asbestos-shingled double barn; collapsing tin barn; white pole barn; beige pole barn. 013 2 7.8

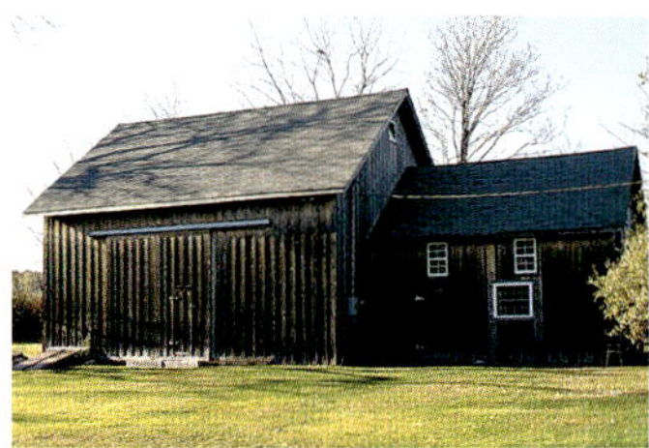

Orient Point, Main Road
Board and batten double barn. The house is dated 1868. While there is no known date on the barn, it is pegged. 013 2 9

Orient Point, Main Road
Gray vertical plank double barn; former cauliflower barn now a dwelling. 018 6 22.1

Orient Point, Main Road
Board and batten barn, new. 018 4 4

Orient Point, Main Road
White horizontal-sided double barn, converted to dwelling. 018 4 3.1

Orient Point, Main Road
Three pegged outbuildings: one vertical plank, two wood-shingled. Victorian house built on cellar of earlier farmhouse. 018 3 17

Orient Point, Main Road
Vertical plank barn once part of preceding farm. 018 3 17, 018 3 30.3

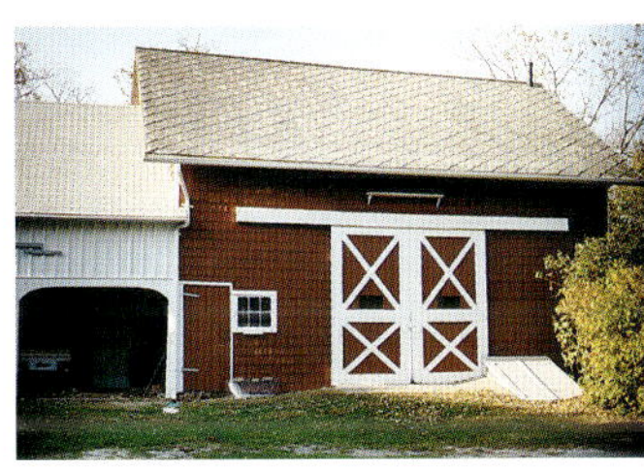

Orient, Main Road
Red wood-shingled barn with white board and batten addition, large stone cellar. The rafters are pinned to one another indicating that the frame is possibly from the eighteenth century. 018 3 14

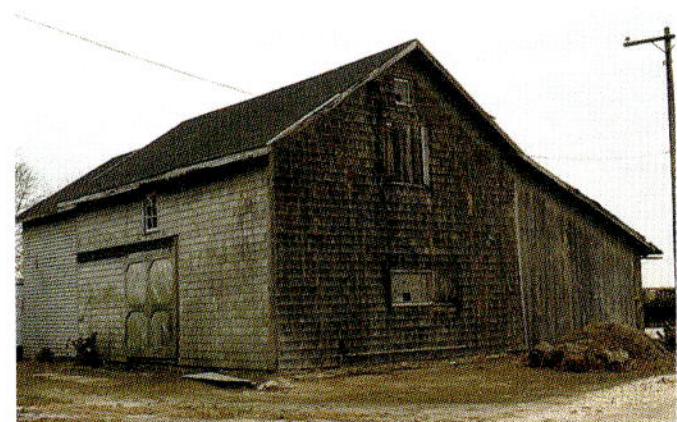

Orient, Main Road
Wood-shingled double barn with shed addition; gray vertical plank double pole barn. 018 6 14.5

Orient, Main Road
Red vertical plank barn; red vertical and horizontal plank English barn with addition circa 1900; red horizontal plank barn with shed addition; red metal barn with addition; new large red barn north behind cluster of barns on the road. House built circa 1800. 018 2 34

Orient, Main Road
Wood-shingled barn, sawn pine frame, former sprout shed. 018 6 3.1

Orient, Platt Road
Vertical plank and board and batten double stable. 018 6 4.1

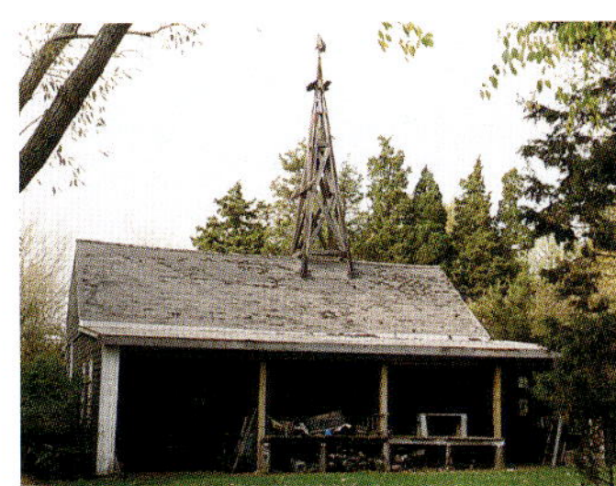

Orient, Main Road
Wood-shingled barn with shed addition built in 1780. A windmill was added atop the barn circa 1894. 018 5 18.3

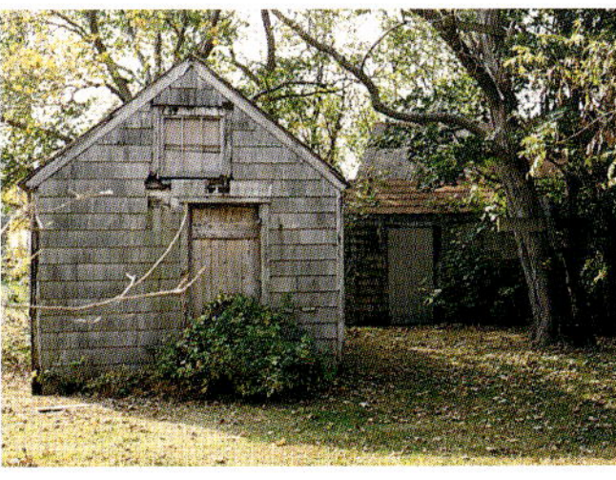

Orient, Platt Road
Wood-shingled hay barn; wood-shingled barn; asbestos-shingled barn. This is what remains of the Hallock farm. 027 2 2.2, 027 2 2.9

Orient, Platt Road
Wood-shingled carriage house with shed additions; two outbuildings. These were also once part of the Hallock farm. 027 1 8

Orient, Halyoke Road
Brown barn; now a dwelling. 027 1 3

Orient, King Street
Red vertical plank barn with addition. 026 2 46

Orient, King Street
Collapsing vertical plank barn. 026 2 45

Orient, King Street
Clapboard and wood-shingled barn. 026 2 43.2

Orient, King Street
Whitewashed vertical plank barn, tin roof. 026 2 37

Orient, King Street
Blue vertical plank and natural wood-shingled double barn with shed addition. Portions of the frame are pegged and there are some joining marks. These elements may come from an earlier barn. 026 2 38 (page 61)

Orient, Willow Terrace
Two large barns, numerous outbuildings. 026 2 39.12

Orient, Willow Terrace
Wood-shingled water tower converted to house. 026 2 39.17

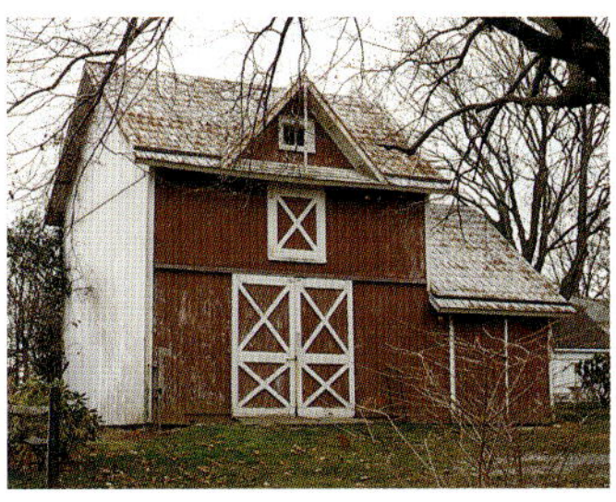

Orient, Navy Street
Red and white vertical plank barn. 026 1 27

Orient, Willow Street
Wood-shingled barn; now a dwelling. 026 1 12

Orient, Navy Street
Wood-shingled and vertical plank carriage house. 026 1 12

Orient, Village Lane
Green vertical plank barn with flanking shed additions; now a dwelling. 025 3 11

Orient, Village Lane
Red vertical plank barn, stone foundation. *Oysterponds Historical Society* 025 3 16.1

Orient, Orchard Street
Red asbestos-shingled barn. 025 4 7

Orient, Orchard Street
Wood-shingled barn with addition. 025 4 8

Orient, Orchard Street
Gray vertical plank barn, wood-shingled roof. 025 2 22.1

Orient, Tabor Road and Orchard Street
Gray vertical plank double barn; gray vertical plank attached sheds; gray barn. 025 2 21.1 (page 107)

Orient, Orchard Street
Red vertical plank double carriage house. 025 2 20.22

Orient, Orchard Street
Red vertical plank and wood-shingled barn with clapboard addition. 025 2 18

Orient, Orchard Street
White vertical plank double barn. 025 2 11.1

Orient, Village Lane
Green vertical plank and asphalt-shingled barn with addition. 025 1 27.1

Orient, Village Lane
Red wood-shingled barn, vertical plank doors. 025 1 18

Orient, Village Lane
Red and gray vertical plank barn with additions. 025 2 5.3_

Orient, Main Road
White vertical plank carriage house. (page 129) 017 4 24

Orient, Main Road
Vertical plank double barn, stone foundation. 017 4 16

Orient, Main Road
Wood-shingled barn; now dwelling. 017 4 15

Orient, Birdseye Road
Two vertical and horizontal plank stables, one converted to a dwelling. 017 2 1.13

Orient, Young's Street
Barn converted to a studio. 018 1 10

Orient, Main Road
Small wood-shingled barn. 017 6 10

East Marion, Main Road
White vertical plank barn with shed additions; small outbuilding. 022 3 38

East Marion, Main Road
Wood-shingled baymen's shack with shed addition. 031 5 10.1

East Marion, Main Road
Wood-shingled baymen's shack with shed additions; clapboard barn with addition. 031 5 9.1

East Marion, Main Road
White vertical plank attached barn with stone foundation; now a dwelling. 031 14 7

East Marion, Main Road
Red vertical plank and wood-shingled double barn; outbuildings. 031 5 1.1 (page 101)

East Marion, Main Road
Wood-shingled barn, at least seventy-five years old. 031 4 31

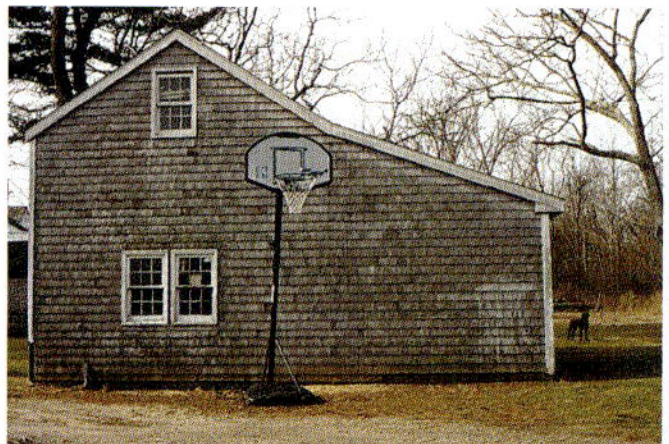

East Marion, Main Road
Gray wood-shingled baymen's shack with shed additions. 031 4 16.6

East Marion, Main Road
Gray vinyl-sided large barn, mortised; original frame intact. 031 3 4.28

East Marion, Main Road
Vertical plank double barn, more than one hundred years old, belonged to Captain Leak; white vertical plank shed. (page 49) 031 3 17

East Marion, Main Road
White clapboard barn. 031 3 16

East Marion, Main Road
White shingled barn; the west end of the barn was a carriage shed. 031 3 12

East Marion, Main Road
Pink vertical plank baymen's shack with additions. The small shed contained an iron well used for tarring fishing nets. (page 63) 031 3 14

East Marion, Main Road
Yellow vertical plank baymen's shack. 031 3 13

East Marion, Main Road
White vertical plank carriage house; white cinder-block chicken coop. 031 2 33

East Marion, Main Road
Red clapboard and vertical plank hay barn with shed addition. 031 2 32.4

East Marion, Rocky Point Road
New barn, a nursery operation. 031 2 21.6

East Marion, Main Road
Gray asphalt- and wood-shingled triple barn with additions. The barns vary in age. The oldest barn is set to the west. 031 1 5.9

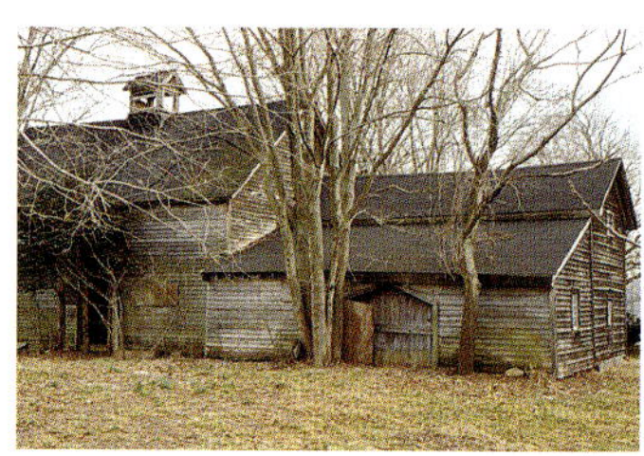

East Marion, Main Road
Horizontal wood plank double barn. The garage behind the house was a carriage house and the white shed was a spotter post during World War II. 031 1 1

East Marion, Main Road
Metal pole barn. 038 1 1.3

East Marion, Main Road
Vertical plank baymen's shack. 031 6 1

East Marion, Main Road
White clapboard baymen's shack; tin-roofed shed. 031 6 8

East Marion, Main Road
Gray vinyl-sided baymen's shack; cupola. 031 6 17.2

East Marion, Old Orchard Lane
Dark gray vertical plank baymen's shack with additions. 031 7 21

East Marion, Main Road
Red vertical plank baymen's shack with shed addition; chicken coop. 031 7 4

East Marion, Main Road
Beige vertical plank barn with addition. 031 8 1.1

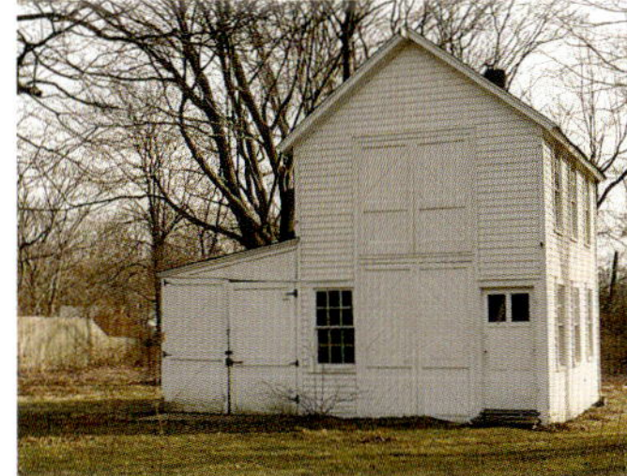

East Marion, Main Road
White clapboard baymen's shack with shed addition, outbuildings. 031 8 2

East Marion, Bay Avenue
Green vinyl clapboard barn, now a dwelling. 031 8 5

East Marion, Bay Avenue
Vertical plank barn with shed addition. 031 10 12

Greenport, Maple Lane
Stucco barn complex, former dairy farm. Two dairy barns, one hay barn with extensions. 035 8 3.2, 035 8 4.1 (page 41)

Greenport, Main Road
White vertical plank barn. 035 2 17.1

Greenport, Main Road
Beige barn complex including hay barn. 035 2 9.2 (pages 90–91)

Greenport, Main Road
Red vertical plank and wood-sided double barn with shed addition; metal pole barn. 035 2 1

Greenport, Main Road
Green asphalt and wood-shingled barn, shed addition. 035 2 4

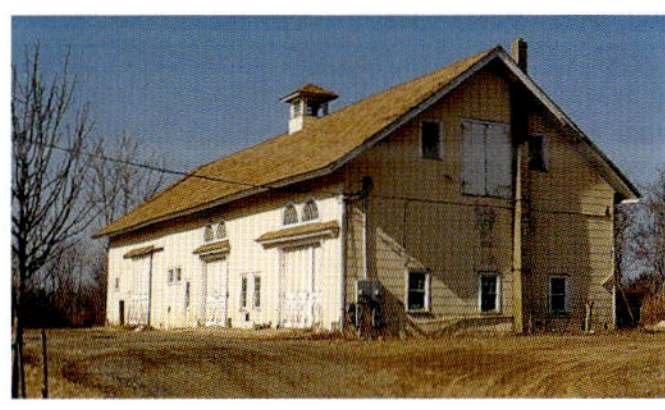

Greenport, Main Road
Yellow asbestos shingle over vertical plank barn. This ornate barn is part of Brecknock Hall, which was the Delafield Robinson house. Delafield Robinson was the grandson of William Floyd. 035 1 25

Greenport, Main Road
Brown cinder-block and wood-shingled chicken barn; now a dwelling. 034 2 2

Greenport, Sound View Avenue
White clapboard gambrel-roofed barn now apartments. 035 1 22

Greenport, Sound View Avenue
White board and batten barn. 033 5 13.1

Greenport, Sound View Avenue
Vertical plank barn, topless silo. 033 3 18

Greenport, Sound View Avenue
Red clapboard barn with cupola; now a dwelling. 040 2 9

Greenport, North Road
Collapsing wood-shingled barn with shed addition, circa 1930, metal pole barn, two small outbuildings. 044 4 4

Greenport, Colony Road
Wood-shingled double barn, more than one hundred years old. Prior to current stable use it was a chicken coop. 052 5 60.2 (page 105)

Greenport, Albertson Lane
Vertical plank barn; Quonset hut barn. 052 4 2

Greenport, Albertson Lane
Red vertical plank barn; red clapboard barn. 052 4 4

Greenport, Main Road
White asbestos-shingled barn; cinder-block and vertical plank gambrel-roofed barn. 053 1 1.2

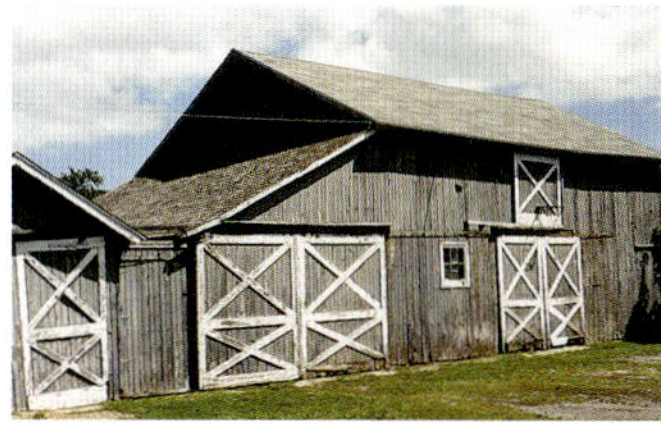

Greenport, Pipes Neck Road
Gray vertical plank barn with two attached structures and shed addition, built around 1920. The early saltbox house is named after James Corwin. 053 1 6

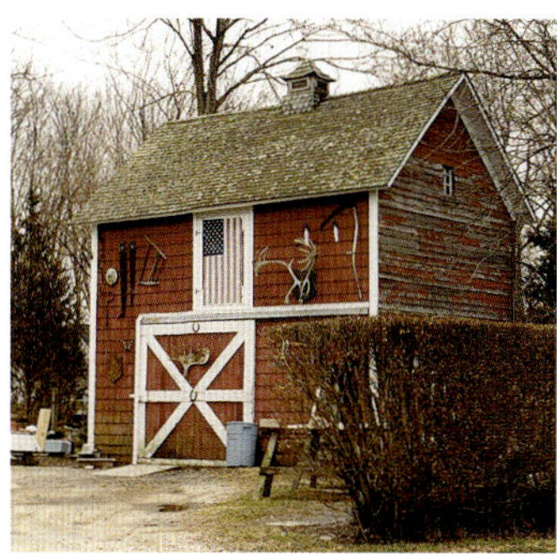

Greenport, Front Street
Red clapboard and wood-shingled carriage house. 048 1 3

Greenport, Front Street
Red wood-shingled carriage house. 048 1 13

Greenport, 5th Street
Brick gambrel-roofed carriage house. 007 2 10

Greenport, 5th Street
White vertical plank carriage house with shed addition. 006 7 9

Greenport, Front Street
Red vertical plank barn. 004 4 28.1

Greenport, 5th Avenue
Red vertical plank carriage house with shed addition. 004 4 37

Greenport, Bay Avenue
Gray wood-shingled barn; now a dwelling. 005 2 1.1

Greenport, Carpenter Street
Yellow stucco double barn. 004 7 29.1

Greenport, Main Street
Vertical plank barn. 004 7 19 (page 29)

Greenport, Center Street
Green carriage house; now a dwelling. 004 2 40

Greenport, Carpenter Street
Vertical plank barn. 003 5 1.2

Greenport, Carpenter Street
Yellow vertical plank carriage house. 004 3 29

Greenport, Sterling Avenue
White clapboard and wood-shingled barn. 003 5 16.5

Greenport, Sterling Avenue
White vertical plank carriage house. 003 4 29

Greenport, Main Street
White vertical plank and clapboard carriage house. 002 6 50

Greenport, Broad Street
Wood-shingled carriage house. 002 5 8

Greenport, 2nd Street
Gray vertical plank barn with cupola. 002 5 23

Greenport, Main Street
White vertical plank and clapboard carriage house with shed addition. 002 1 24

Greenport, Main Street
White vertical plank carriage house with shed addition and wood-shingled roof. 002 1 16

Southold, Main Road
Red vertical plank barn with brick foundation. 056 4 20.1

Southold, Main Road
White wood-shingled carriage house with attached structures. 056 4 15

Southold, Main Road
Two formerly attached wood-shingled barns, both more than one hundred years old; small vertical plank and wood-shingled barn built in the 1930s. 056 6 11.1 (page 89)

Southold, Main Road
Red potato barn; now a dwelling. 056 3 12

Southold, Main Road
Barn complex containing two red wood-shingled barns (a single and a double barn) and a white stucco barn. The eastern portion of the double barn was built in the eighteenth century. The house dates to the early eighteenth century. 056 3 9 (pages 68–69)

Southold, Laurel Avenue
Wood-shingled potato barn. 056 3 2.3

Southold, Laurel Avenue
Wood-shingled barn with flanking shed additions. 056 2 6.1

Southold, Main Road
Red clapboard barn, circa 1870, resting on the original stone foundation. It was extended for use as a milking barn in 1921. Outbuildings include: carriage shed, contiguous stables, chicken coops (1860–70), rooster stalls, corncrib, brick ice house. 056 5 1.3 (pages 134–35)

Southold, Main Road
Red vertical plank double barn; metal pole barn. 066 2 2.1

Southold, Main Road
Wood-shingled barn with shed addition. 055 6 33.2

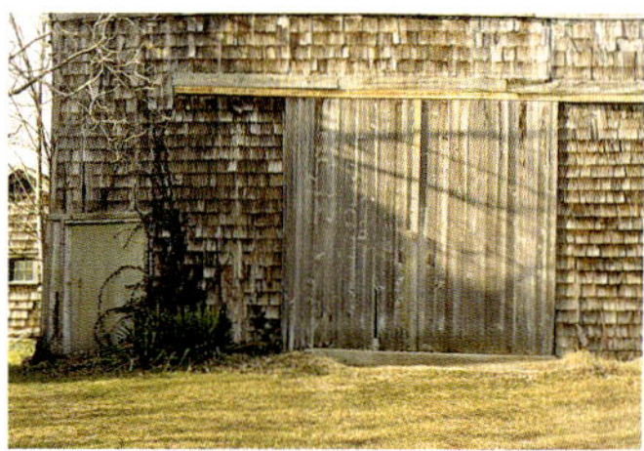

Southold, Main Road
Wood-shingled barn with shed addition; white vertical plank barn with shed addition; wood-shingled barn; two additional wood-shingled structures. 063 3 25

Southold, Main Road
Red vertical plank double barn built in the mid-nineteenth century; three additional outbuildings. 063 3 18.1 (page 15)

Southold, Main Road
Wood-shingled double barn, formerly a stable. 063 3 28.1

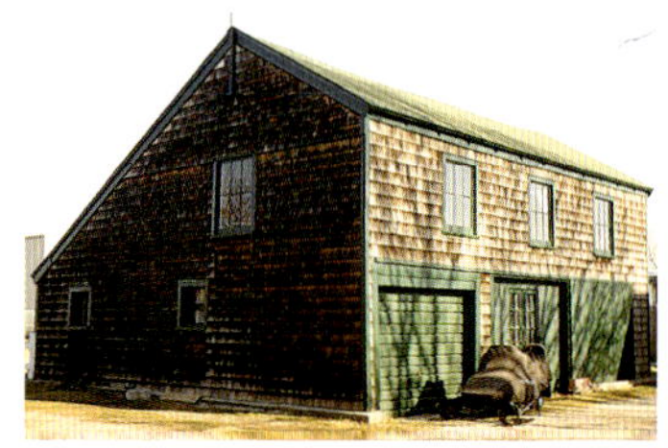

Southold, Main Road
Wood-shingled stable. The saltbox form is achieved by a shed addition on the north side. Two layers of vertical plank siding remain intact under the wood shingles. The structure was built and used as a stable. 063 3 10

Southold, Route 48
Vertical plank barn with shed addition. 055 3 4.4 (pages 102–03)

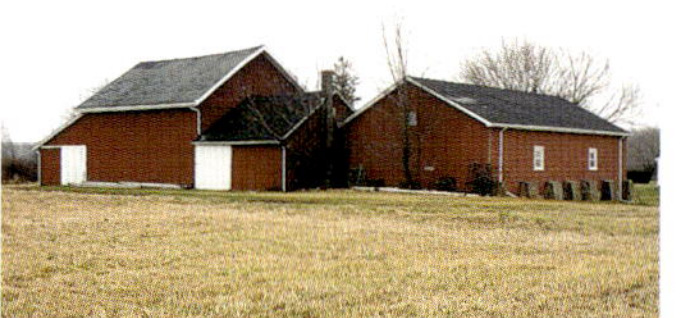

Southold, North Road
Red wood-shingled and vertical plank double barn with shed addition; two additional outbuildings. 051 3 4.3

Southold, North Road
Cinder-block/stucco double barn. 051 3 5

Southold, Mount Beulah Avenue
Brick carriage house; now a dwelling. 051 3 2.10

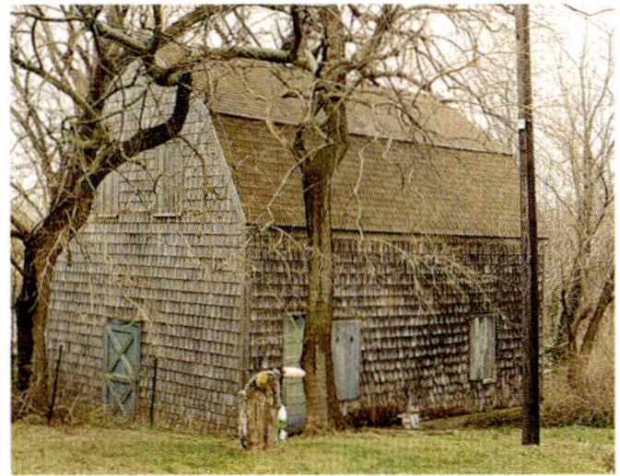

Southold, Sound View Avenue
Wood-shingled gambrel-roofed barn, dates to nineteenth century, formerly part of the Clark farm. At one time there was a windmill. 135 1 3

Southold, Sound View Avenue
Wood-shingled three-story carriage house, part of the Dr. Marshall estate, built in 1915; new vertical plank stable. 050 2 15, 051 1 1

Southold, Route 48
Two red vertical plank barns, circa 1920, one with flanking shed additions. 055 2 10

Southold, Boisseau Avenue
White vertical plank barn now used as stable; gray wood-shingled barn; small wood-shingled barn. 055 5 12.1

Southold, Boisseau Avenue
Vertical plank barn with shed addition, water tower. 055 6 8

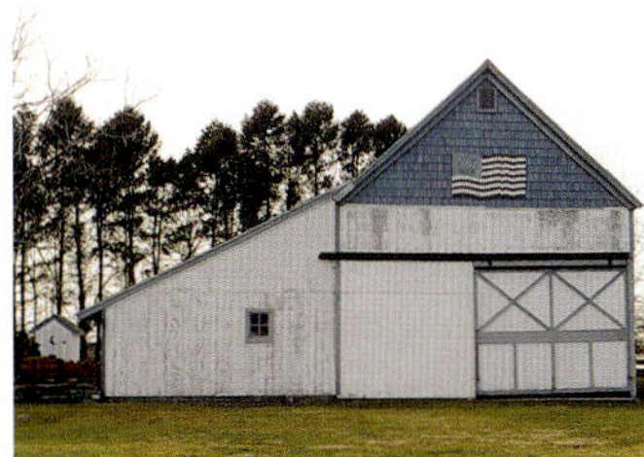

Southold, Boisseau Avenue
White vertical plank barn; white vertical plank and blue wood-shingled barn with shed addition. The dates of construction are not known, possibly the mid-nineteenth century. 055 5 16

Southold, Boisseau Avenue
Vertical plank barn. 063 3 7

Southold, Youngs Avenue
Gray vertical plank barn. 060 2 8

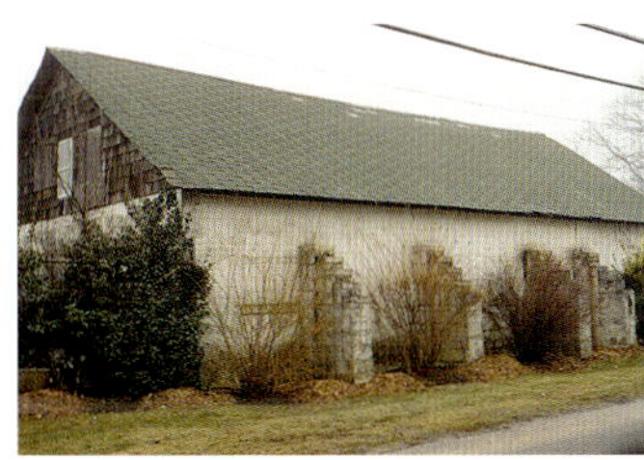

Southold, Hummel Avenue
Buttressed stucco former potato barn. 063 2 30.1

Southold, Hummel Avenue
Asbestos-shingled barn built circa 1900. 063 2 22

Southold, Youngs Avenue
Beneath the white vinyl siding is an "old" barn. 063 2 5

Southold, Youngs Avenue
Wood-shingled barn; gray asphalt- and wood-shingled barn; red vertical plank barn; metal Quonset hut barn. 055 2 9.4 (page 83)

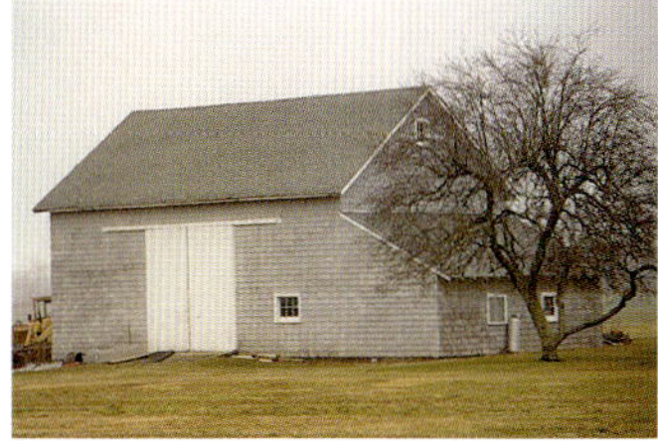

Southold, Youngs Avenue
Gray wood-shingled barn with shed addition; gray vertical plank barn; outbuildings. 055 2 8.5

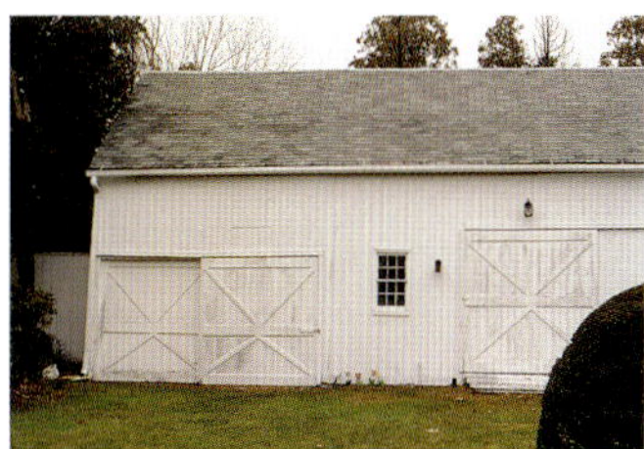

Southold, Lighthouse Road
White vertical plank and board and batten barn with addition. 055 1 2

Southold, Sound View Avenue
Red metal pole barn. 050 5 1

Southold, Lighthouse Road
White wood-shingled barn. 050 1 11

Southold, North Road
Wood-shingled and vertical plank barn complex, with several additional outbuildings. 054 3 24.1

Southold, North Road
Wood-shingled barn. 055 1 7

Southold, Horton Lane
Wood-shingled and vertical plank double barn with shed additions; vertical plank raised barn built in the nineteenth century; additional outbuildings. 054 7 21.1 (page 127)

Southold, Middle Road
Green metal-sided barn, former site of Long Island Cauliflower Association auctions. 055 1 11.4

Southold, Horton Lane
Vertical plank gambrel-roofed barn with addition; vertical plank carriage house. 063 1 3.1

Southold, Horton Lane
Red vertical plank and wood-shingled barn. 063 1 8.1

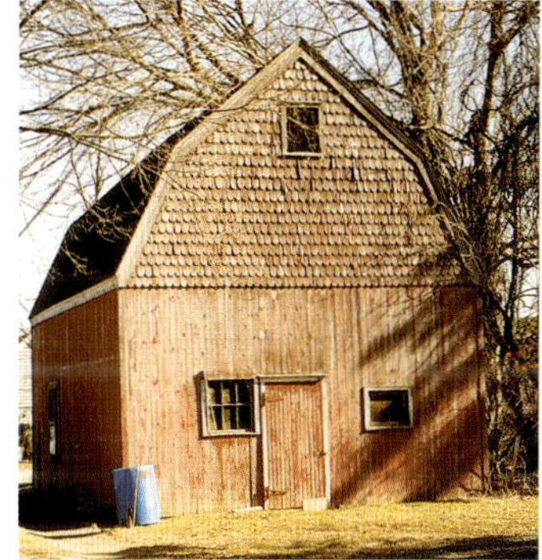

Southold, Main Road
Red vertical plank and sculpted wood-shingled barn. 061 1 9.1

Southold, Main Road
Gray shingled carriage house; now a dwelling. House built in 1905. 061 1 8.1

Southold, Traveler Street
White vertical plank carriage house. 061 2 2

Southold, Beckwith Street
Yellow vertical plank carriage house. 061 1 24

Southold, Beckwith Street
Brown wood-shingled carriage house built 1870–75. 061 1 23

Southold, Beckwith Street
Pink vertical plank carriage house. 061 1 22

Southold, Main Road
Vertical plank barn. 061 2 7.2

Southold, Youngs Avenue
Vertical plank double barn. 063 1 25

Tuckers Lane
Square wood-shingled carriage house. 063 5 2(7)

Southold, Tuckers Lane
Square white stucco carriage house. 063 5 2(6)

Southold, Tuckers Lane
Wood-shingled carriage house. 063 5 2(5)

Southold, Tuckers Lane
Board and batten carriage house. 063 5 2(8)

Southold, Tuckers Lane
Red vertical plank barn; three outbuildings. 059 11 5

Southold, Horton Lane
Red vertical plank barn; square metal pole barn; three connected barns. 063 1 1.3

Southold, Route 48
Vertical plank and wood-shingled barn. 059 4 2.1

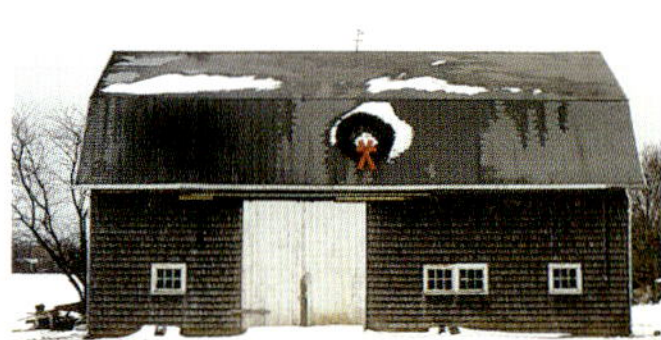

Southold, Route 48
Wood-shingled gambrel-roofed barn with shed addition. 059 4 2.3

Southold, Route 48
Gray vertical plank barn with flanking wood-shingled shed additions. 059 7 30

Southold, Route 48
Vertical plank barn. 059 9 30.8

Southold, North Road
Vertical plank barn; wood-shingled barn; square wood-shingled carriage house; two sheds. 069 2 1

Southold, Ackerly Pond Lane
Red vertical plank and wood-shingled triple barn; red vertical plank barn; carriage house; greenhouses and outbuildings. 069 2 3

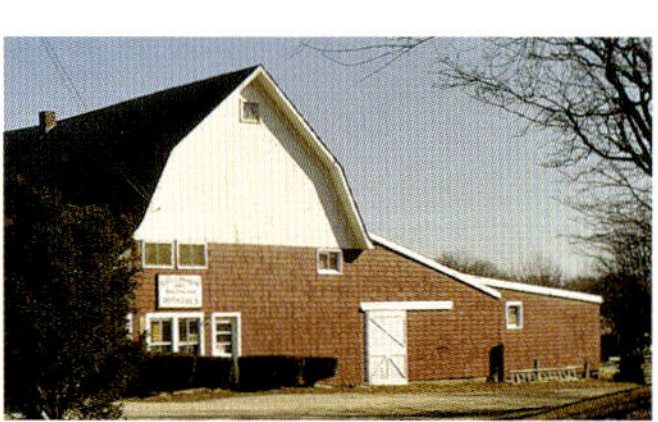

Southold, North Road
Red wood-shingled gambrel-roofed barn with shed additions; metal pole barn; worker cottages. 069 3 2

Southold, Ackerly Pond Lane
White metal pole barn. 069 5 4.1

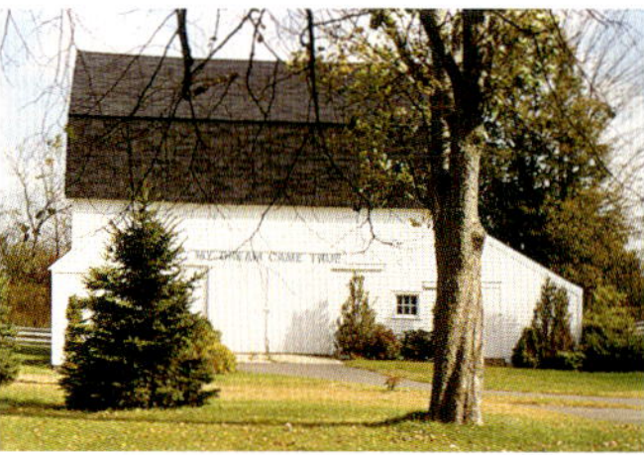

Southold, Ackerly Pond Lane
White clapboard barn with shed addition; three sheds. 069 3 9.2

Southold, Ackerly Pond Lane
Red vertical plank barn with shed additions. 069 5 7.1 (page 155)

Southold, Ackerly Pond Lane
Small vertical plank barn. 069 5 12.1

Southold, Main Road
Plywood-covered collapsing barn. 069 3 10.1

Southold, Main Road
White asbestos-shingled carriage house. 069 6 6

Southold, Main Road
White vinyl clapboard and wood-shingled gambrel-roofed barn. 069 6 7

Southold, Main Road
White vertical plank carriage house. 070 7 10

Southold, Main Road
Gray vertical plank barn. 070 7 1

Southold, Main Road
Red vertical plank tin-roofed barn. 070 1 3

Southold, Main Road
Yellow vertical plank barn. 070 1 12

Southold, Main Road
Wood-shingled gambrel-roofed barn. 063 6 2

Southold, Main Road
White vertical plank barn. 070 2 4

Southold, Main Road
Gray vertical plank barn with shed addition. 070 2 1

Southold, Main Road
Small red vertical plank barn. 063 6 3

Southold, Main Road
Gray and red vertical plank double barn; smaller barn with shed addition alongside. English three-bay barn, hand-hewn timbers, mortise and tenon joints, built in the nineteenth century or possibly earlier. Second barn, to the east, contains dates from the mid-nineteenth century. The house is the Moses Cleveland house. 063 6 8 (pages 56–57)

Southold, Wells Avenue
Asbestos-shingled barn, brick cellar partially exposed. 061 3 8.5 (page 123)

Southold, Calves Neck Road
Red shingled triple—possibly quadruple—barn with addition. 063 7 39

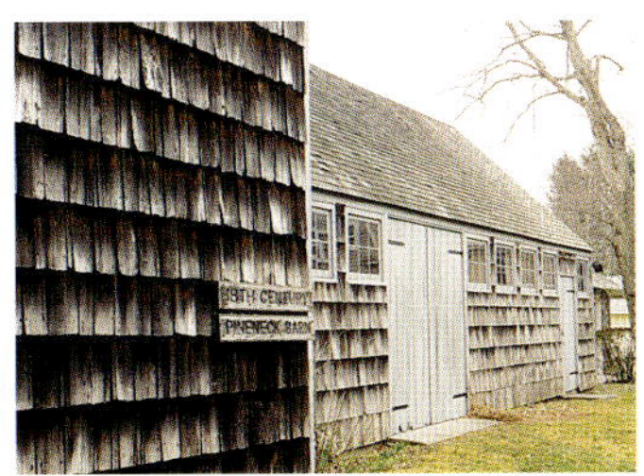

Southold, Main Road
Wood-shingled double English barn circa 1800; red vertical plank carriage house; corncrib; brick ice house. The barn was moved to the site from Pine Neck. *Southold Historical Society* 062 2 5.2

Southold, Hobart Avenue
Wood-shingled carriage house. 064 3 7

Southold, Hobart Avenue
Wood-shingled gambrel-roofed barn. 064 3 3

Southold, Hobart Avenue
Red wood-shingled barn. 064 2 12

Southold, Maple Avenue
Brick carriage house built 1910–15; wood-shingled chicken coop. 064 1 29

Southold, Towns Harbor Lane
White vertical plank barn; smaller vertical plank barn; three outbuildings. The original barn burned down in 1918–19 and was rebuilt in 1920. 063 4 11 (page 151)

Bayview, Pine Neck Road
Vertical plank and wood-shingled barn with shed addition. 070 5 28

Bayview, Pine Neck Road
Red vertical plank carriage house. 070 5 45

Bayview, Custer Avenue
Wood-shingled gambrel-roofed barn; now a home. 070 9 15

Bayview, Bayview Avenue
Board and batten English hay barn built in the late nineteenth century; two sheds, same period. House built in the 1770s. 070 7 19 (page 147)

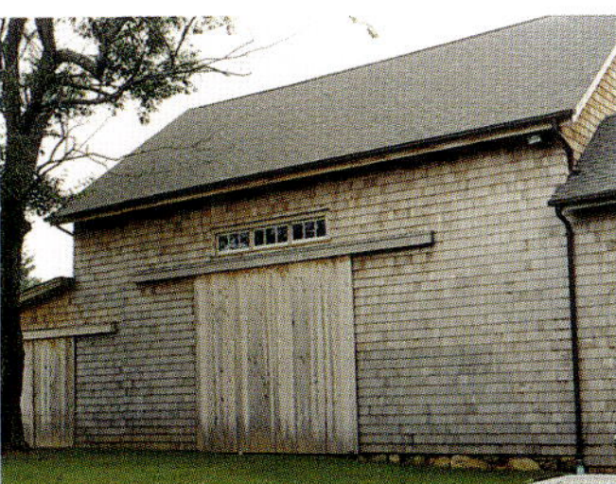

Bayview, Bayview Avenue
Wood-shingled double English barn with shed additions. The house was built in 1753 and the English barn was probably built at the same time. A careful restoration of the barn and its shed additions preserved the roof pitch and an original interior corncrib. 078 1 10.23

Bayview, Main Bayview Road
Vertical plank barn with shed additions built in the nineteenth century. Basic English hay barn configuration extended on the southern end by 6 to 8 feet. 075 4 22.3 (page 33)

Bayview, South Harbor Road
Wood-shingled English barn with wood-shingled wing converted to carriage house. 075 7 1.4

Bayview, South Harbor Road
White asbestos-shingled triple barn built in the nineteenth century; four barns/outbuildings in farm use. 075 7 2

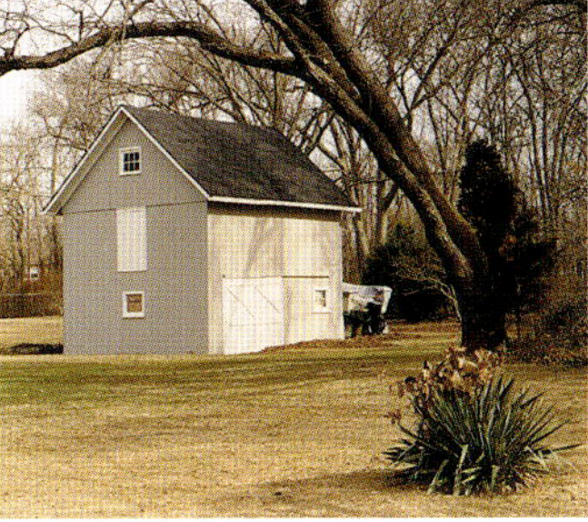

Bayview, Nokomis Road
Gray vertical plank carriage house. 078 3 24

Bayview, Main Bayview Road
Small vertical plank barn. 076 1 8

Bayview, Main Bayview Road
Gray vertical plank gambrel-roofed barn. 078 7 32.12

Bayview, Main Bayview Road
Small horizontal plank barn. 078 9 48

Bayview, Main Bayview Road
Red vertical plank barn moved from “off Carpenter Street” in Greenport. 078 9 61 (page 99)

Bayview, Jacobs Lane
Original owner was Silas Austin Horton Dayton. The Dayton home and carriage house are now on adjoining lots. Asphalt-shingled English barn with shed addition was moved onto the property in 1900 and later converted for potato storage. Three additional barns and vertical plank corncrib. 088 1 10

Bayview, Main Bayview Road
Vertical plank carriage house with cupola and wood-shingled roof, originally Dayton. 088 1 7 (page 115)

Bayview, Main Bayview Road
White cinder-block barn. 088 1 8

Bayview, Main Bayview Road
Wood-shingled barn; horizontal plank shed. 088 1 11, 088 1 11

Bayview, Main Bayview Road
White wood-shingled carriage house. 088 1 12

Bayview, Midland Parkway
Wood-shingled barn moved from across the street; now a dwelling. 088 2 12.1

Bayview, South Harbor Road
Local terra-cotta brick gambrel-roofed barn built in 1908 by Haven Emerson; red vertical plank three-bay English hay barn with addition built between 1800 and 1850; two additional red vertical plank barns circa 1900. The vertical plank barns were moved back from the farmhouse in the early twentieth century. 086 3 1 (page 27)

Cutchogue, Main Road
Brown wood-shingled carriage house. 085 2 13.1

Cutchogue, Main Road
Gray and green asphalt-shingled double barn. 085 2 12.3

Cutchogue, Bay Avenue
Gray vertical plank barn with addition; metal pole barn. 085 3 1

Cutchogue, Main Road
Wood-shingled English barn with addition, altered for potato storage in 1919; board and batten attached barn built 2000; board and batten barn also built in 2000. Farm began with a land grant to the Hutchinson family in 1724. The house was built in 1750. 097 1 25.1, 085 2 10.3

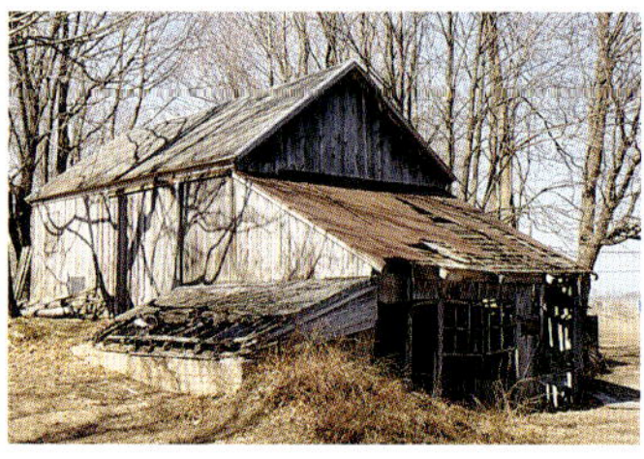

Cutchogue, Main Road
Vertical plank barn with shed additions. Converting barns to homes or businesses happens routinely. The converse is rarely true. The Old House in Cutchogue was used as a barn for a brief period in its long life. This barn was a general store on the Main Road in the nineteenth century. It was moved south on the property in the early twentieth century and converted to a barn for a nursery operation specializing in flowers. 097 3 3

Cutchogue, Main Road
Gray vertical plank gambrel-roofed stable. 097 3 1

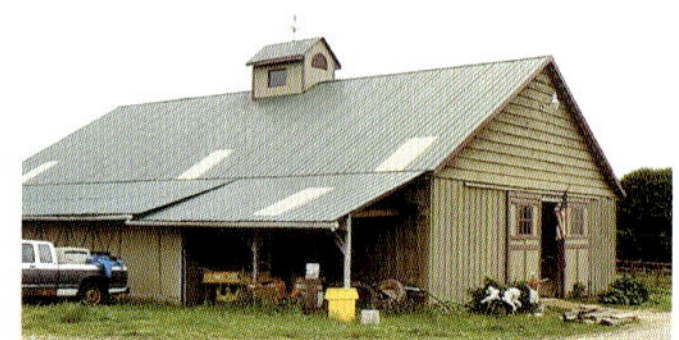

Cutchogue, Main Road
Beige board and batten stable with shed addition. 097 2 23

Cutchogue, Leslie's Road
Vertical plank barn built in 1934, recently restored using original wood. 097 4 19

Cutchogue, Bay Avenue
Vertical plank and board and batten barn. 097 4 9

Cutchogue, Bay Avenue
Vertical plank double barn with shed addition. 097 4 11

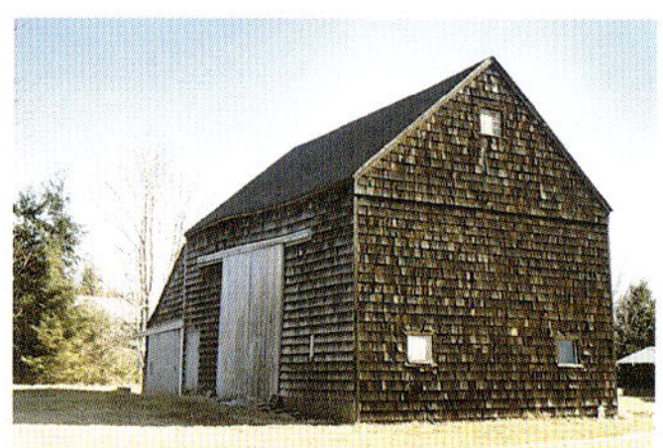

Cutchogue, Bay Avenue
Wood-shingled English barn with shed addition built in the late nineteenth century; carriage house; three outbuildings. 097 9 9

Cutchogue, Bay Avenue
Gray corrugated-tin barn. 097 8 31.4

Cutchogue, Bay Avenue
White square-shingled barn; two gray asphalt-shingled structures. 097 9 10.3

Cutchogue, Bay Avenue
Light gray vertical plank barn built circa 1900. 097 9 13

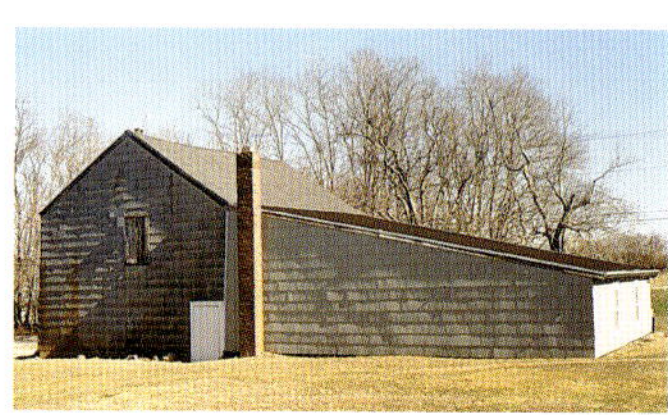

Cutchogue, Eugene's Road
Two-story asbestos-shingled English barn with elongated lean-to addition, pinned, built in the late nineteenth century; gray asbestos-shingled barn. 097 7 27

Cutchogue, Main Road
Wood- and asbestos-shingled barn with flanking shed additions; yellow pole barn. 097 1 12.6, 097 1 12.7, 097 1 12.8

Cutchogue, Main Road
Wood- and gray asbestos–shingled barn, now a carpenter's shop. 097 1 14

Cutchogue, Main Road
Gray asbestos-shingled potato barn; wood-shingled barn. 097 1 11.4

Cutchogue, Main Road
Wood-shingled barn with shed addition. 097 2 12.1

Cutchogue, Cox Lane
White board and batten stable. 097 1 1

Cutchogue, Cox Lane
Small red vertical plank barn. 096 3 8

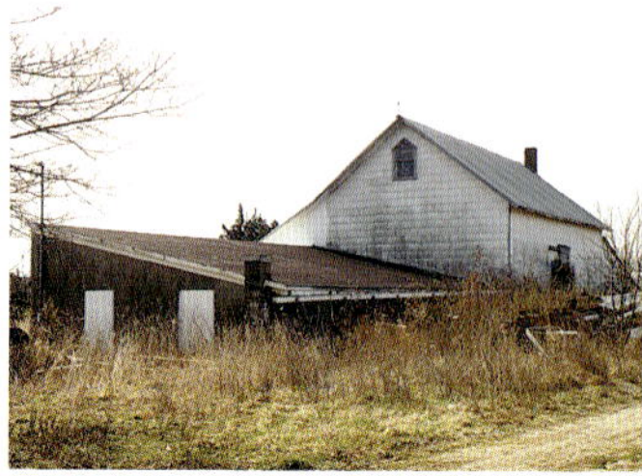

Cutchogue, Cox Lane
Gray asbestos-shingled barrel barn; white vertical plank and shingled barn with additions. 097 5 2.1

Cutchogue, Cox Lane
White shingled barn with addition. 097 5 2.2

Cutchogue, Main Road
Wood-shingled barn with shed addition built late nineteenth or early twentieth century; wood-shingled seed-sorting barn with potato cellar. The house is more than two hundred years old. 097 6 7

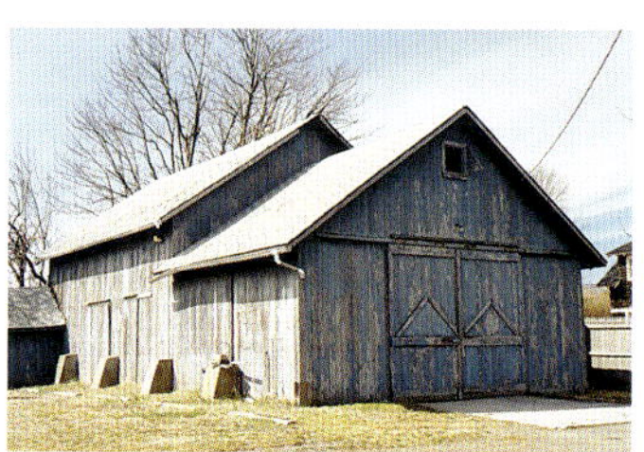

Cutchogue, Main Road
Blue vertical plank double barn with small buttresses. 097 6 15

Cutchogue, Main Road
Gray vinyl clapboard double barn. The exterior is deceptive; the framing reveals that these are old barns. 097 5 6

Cutchogue, Main Road
Two reddish brown vertical plank barns; large beige double metal pole barn; greenhouses. 096 3 9

Cutchogue, Main Road
Red barn complex consisting of two barns, a shed, and a carriage house. Materials include vertical plank and wood shingles. 103 1 19.2 (page 35)

Cutchogue, Main Road
Red vertical plank barn with shed addition; green shingled potato barn. 102 2 24.2

Cutchogue, Main Road
Gray/green vertical plank barn with shed addition. 102 2 18

Cutchogue, Main Road
Vertical plank five bent gambrel-roofed barn with attached shed built in the mid-nineteenth century. This estate was originally Wells. Fleet married into the family and raised racehorses around the time this barn was constructed. Renovated in 2003. 102 3 10

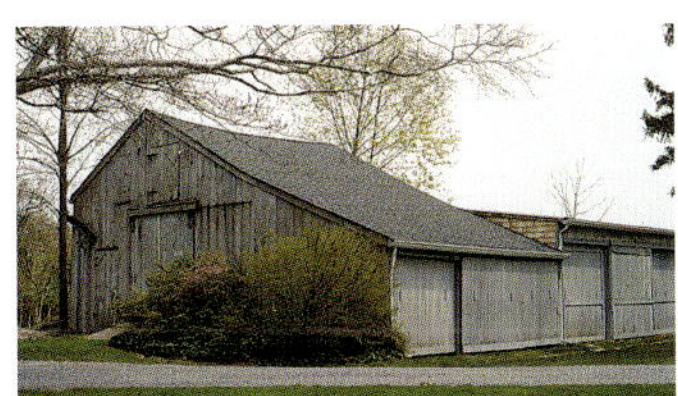

Cutchogue, Main Road
Vertical plank and wood-shingled stable with addition. This was attached to two English barns and a carriage shed (surveyed in 1970; now gone) that were part of the Fleet barn complex. Evidence of the carriage shed appears on the north side of the stable. 102 6 23.1

Cutchogue, Depot Lane
Small vertical plank barn with shed addition, not original to the site. 102 2 12.1

Cutchogue, North Street
Red vertical and horizontal plank barn with shed addition; currently an art gallery. 102 5 18

Cutchogue, Griffin Street
Square brown wood-shingled and white clapboard carriage house with cupola. 102 5 17

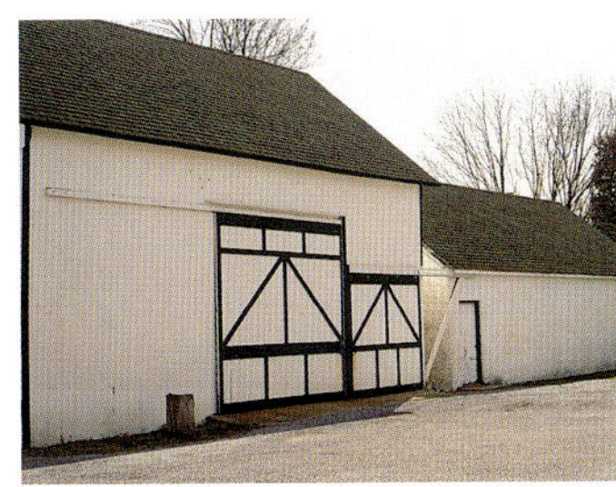

Cutchogue, Main Street
White vertical plank barn; adjacent building. 102 6 3

Cutchogue, New Suffolk Road
White wood-shingled and vertical plank double barn; now a dwelling. 102 6 9

Cutchogue, Main Road
Nine bent gray asbestos-shingled barn built in 1877, with brick cellar access on hillside; asbestos-shingled barn with shed addition built on the foundation of an earlier barn that burned in 1883. The house was built in 1780 and enlarged in the 1820s. 102 6 20.1

Cutchogue, New Suffolk Road
Board and batten barn; now a dwelling. 109 7 4.4

Cutchogue, New Suffolk Road
Gray vertical plank barn with shed additions. 109 6 4

Cutchogue, New Suffolk Road
Gray corrugated-metal barn. 109 6 5

Cutchogue, New Suffolk Road
Board and batten barn; now a dwelling. 109 7 3

Cutchogue, Case's Lane
Two English barns, both collapsing. William Harrison Case in the 1873 atlas. 109 6 9.1

Cutchogue, New Suffolk Road
Wood-shingled and vertical plank English-style barn with additions. This unusual barn is believed to have been built in 1852. It contains timbers from the 1732 Cutchogue Presbyterian church, which was dismantled in 1850. White oak knee braces flank the center bay, the east wall contains a reused wall girt and a section of a summer beam, and there are notched and pegged roof rafters used in the construction. Also: vertical plank barn; additional outbuildings. 109 7 10.3 (pages 110–11)

Cutchogue, Case's Lane
Red board and batten and vertical plank double barn. *Cutchogue Historical Society*. 109 5 6.1

Cutchogue, Main Road
Red board and batten and vertical plank barn; three outbuildings. 109 5 3

Cutchogue, Main Road
Red vertical plank carriage house. 109 2 13.6

Cutchogue, Crown Land Lane
Gray vertical plank modern house incorporating barn; outbuilding. 109 2 12.9

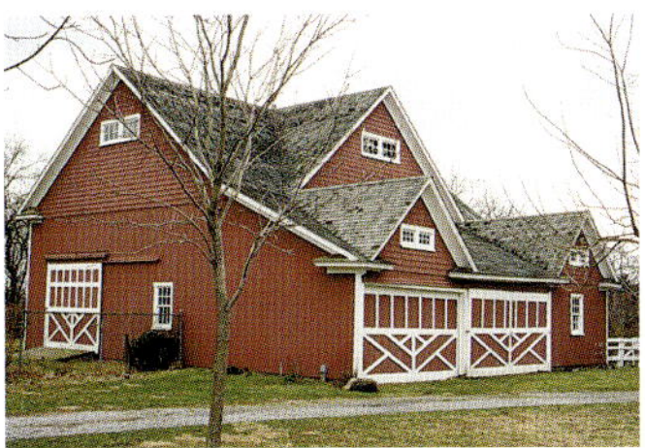

Cutchogue, Moore's Lane
Red vertical plank and wood-shingled stable with additions; red vertical plank outbuilding. Both structures have wood-shingled roofs. 109 4 2

Cutchogue, Main Road
Vertical plank and red asbestos-shingled barn with addition. 109 3 2.3

Cutchogue, Alvah's Lane
Red vertical plank barn with extension, converted to stable use; stucco potato barn; gray vertical plank feed shed; wood-shingled chicken coop. 109 1 24.3

Cutchogue, Main Road
Gray gambrel-roofed vertical plank barn with addition; two red metal pole barns. 109 1 13

Cutchogue, Main Road
Red square vertical plank and wood-shingled barn with shed addition, three-bay English, pegged, dates to eighteenth century; vertical plank and board and batten double barn built in the nineteenth century; cinder-block potato barn built in the twentieth century. 109 1 11 (pages 46–47)

Cutchogue, Route 48
Red vertical plank barn with shed addition. 101 2 6

Cutchogue, Main Road
Gray corrugated metal–covered triple barn. 109 1 10.1

Cutchogue, Main Road
Gray board and batten and wood-shingled barn completed in 1992; board and batten barn with shed addition built in the 1980s. 109 1 8.7

Cutchogue, Main Road
Stucco and gray asbestos-shingled barn. 109 1 40

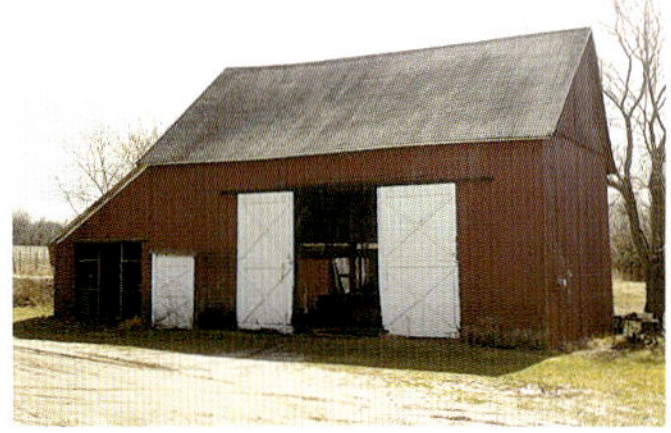

Cutchogue, Main Road
Eighteenth-century tongue-and-groove vertical plank barn. The stone foundation is preserved. A shed addition shelters original Long Island "long" wood shingles on east exterior wall. Late seventeenth– or early eighteenth–century roof purlins have been recycled as studs. *Peconic Land Trust.* 109 1 39

Cutchogue, Main Road
Double vertical plank barn on brick foundation; green asphalt-shingled related farm structures. 109 1 7

Cutchogue, Main Road
Two gray asphalt and asbestos–shingled double barns. 108 3 9.3

Cutchogue, Main Road
Gray vertical plank barn with shed addition; three outbuildings. 115 8 3.3 (page 21)

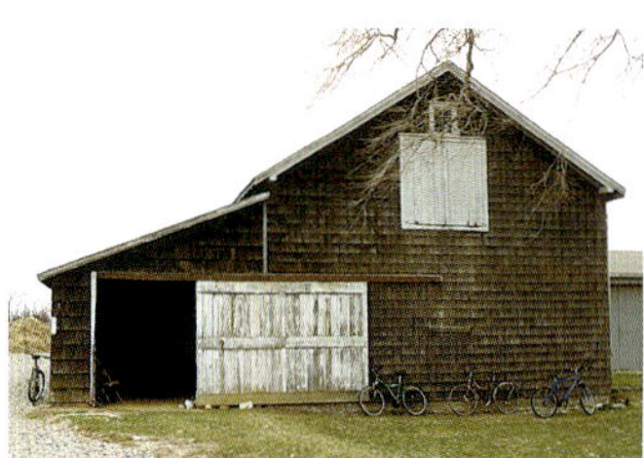

Cutchogue, Route 48
Wood-shingled double barn built in the nineteenth century; brick buttressed potato barn from the eighteenth century; small white former asparagus shed converted for wine storage. 108 3 7

Cutchogue, Alvah's Lane
Double barn; chicken coop. The asphalt-shingled barn was built in the early eighteenth century with a shed addition in 1890. The vertical plank barn was built in 1910. 101 2 22 (pages 78–79)

Cutchogue, Alvah's Lane
Wood-shingled barn with shed addition; wood-shingled barn. The main barn was originally half as wide, split and augmented in the middle. The second barn was probably lifted onto its cinder-block foundation. 101 2 24.5

Cutchogue, Alvah's Lane
Vertical plank barn, disintegrating but braced. 102 4 5.1

Cutchogue, Route 48
Yellow wood-shingled potato barn with flanking shed additions built in 1949; wood-shingled double barn built in 1850 with underground Prohibition-era room; four wood-shingled outbuildings including stable and chicken coop. 101 2 18.3

Cutchogue, Route 48
Metal pole barn with addition, recent construction. 096 4 4.3

Cutchogue, Depot Lane
Potato processing building. 096 4 8.1

Cutchogue, Depot Lane
Four potato barns (gray cinder-block, yellow metal, yellow horizontal vinyl siding, gray double with shed addition). 096 5 1.2

Cutchogue, Depot Lane
Wood-shingled early barn with addition, original foundation. 102 1 6.1

Cutchogue, Depot Lane
Yellow vertical plank barn. 102 1 5.2 (page 75)

Cutchogue, Depot Lane
Three red vertical plank barns. 102 2 2.3

Cutchogue, Depot Lane
Stucco and vertical plank gambrel-roofed barn with shed addition. The barn was built in the 1920s along the railroad tracks. It is a potato barn constructed with steel beams. Potatoes were conveyed to the second floor. Three railroad cars at a time could be loaded from the barn. Corrugated-metal barn on cinder-block stilts; potato processing plant; worker housing. 096 2 3 (page 53)

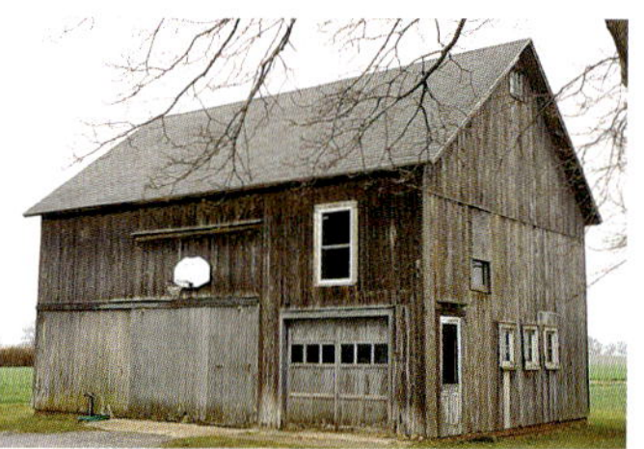

Cutchogue, Depot Lane
Vertical plank barn; asphalt-shingled barn. The main barn is at least eighty years old. 096 2 2

Cutchogue, Route 48
Red asbestos-shingled barn. 096 2 7

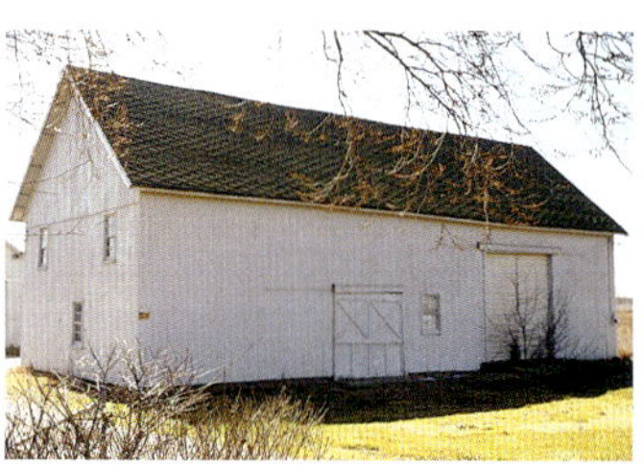

Cutchogue, Route 48
Gray vertical plank barn with addition, pegged, about 130 years old; gray vertical plank barn with addition; gray carriage house. 084 4 1

Cutchogue, Cox Lane
White vinyl clapboard-sided double barn. 096 2 10

Cutchogue, Cox Lane
White vertical planked barn moved to the property; white carriage house; additional outbuildings. 096 3 4.1

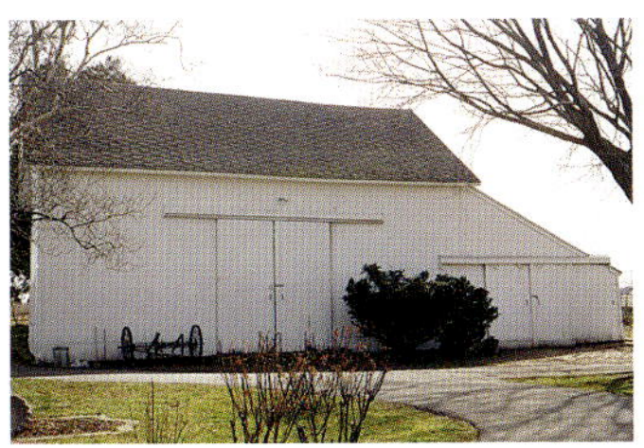

Cutchogue, Cox Lane
White vertical planked barn with shed addition; outbuilding collection. 096 3 3.1

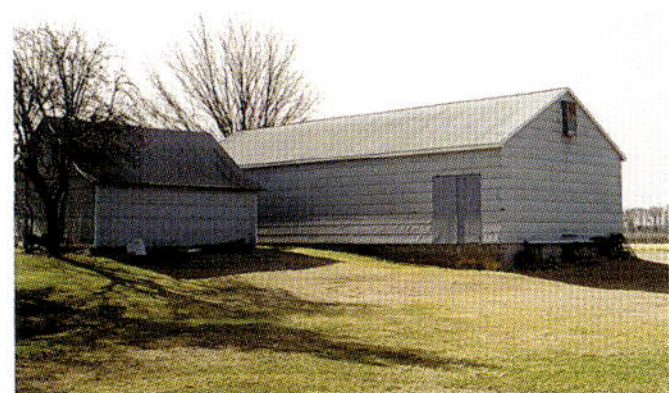

Cutchogue, Bridge Lane
Asphalt-shingled barn; four outbuildings. 085 2 1.1

Cutchogue, Bridge Lane
Vertical plank double barn with addition. 085 2 3

Cutchogue, Route 48
Green asphalt-shingled double barn; outbuildings. 084 5 4 (page 77)

Cutchogue, Route 48
Two beige metal pole barns. 084 4 7.1

Cutchogue, Route 48
Wood-shingled barn with shed addition, pegged. 084 1 9

Cutchogue, Route 48
Metal pole barn. 084 1 25.3

Cutchogue, Cox Lane
Metal pole barn. 084 1 24

New Suffolk, Old Harbor Road
Brown wood-shingled gambrel-roofed barn; now a dwelling. One of twin barns on the Hathaway estate. 117 3 4.2

New Suffolk, Old Harbor Road
Stable from the Hathaway estate enclosed in wood-shingled garage framework. The stable retains wood-shingled siding and wood-shingled roof. Home built in 1873 and was the estate superintendent's house. 117 3 11.5

New Suffolk, Old Harbor Road
Former barn on the Hathaway estate; now a dwelling. 117 3 10

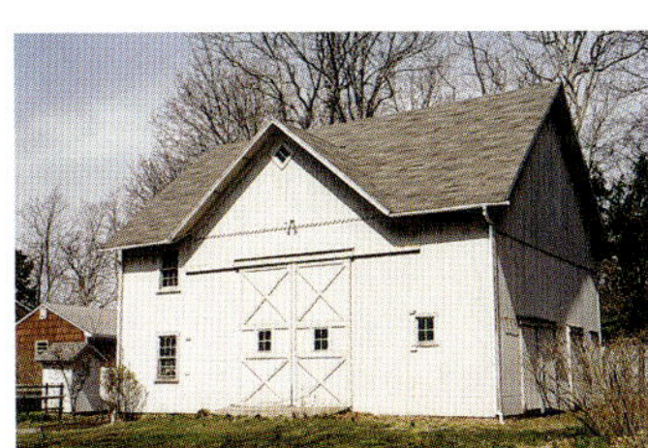

New Suffolk, New Suffolk Road
White vertical plank stable, rock and brick cellar, built prior to the Civil War. 117 5 6

New Suffolk, Old Harbor Road
Vertical plank saltbox structure, Hathaway estate outbuilding. 117 5 7

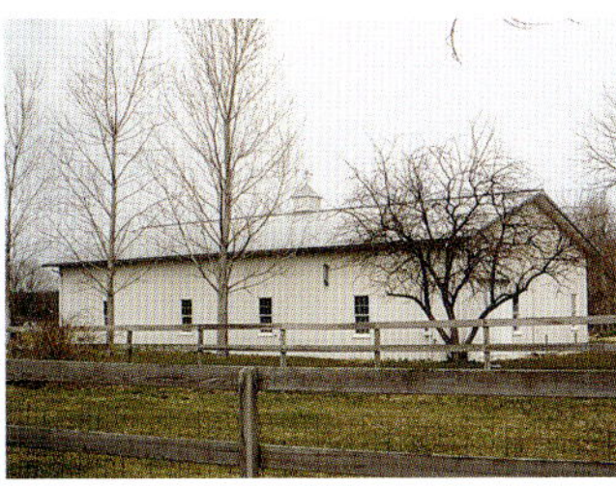

New Suffolk, New Suffolk Road
White metal pole barn; two sheds. Formerly G. Fred Grathwhol dairy farm. 117 2 17.5

New Suffolk, George Road
Red vertical plank barn built 1993–94, with wood-shingled roof, designed by Samuels & Steelman. 117 4 18.4

New Suffolk, New Suffolk Road
Vertical plank barn, with brick foundation. 117 6 7.1

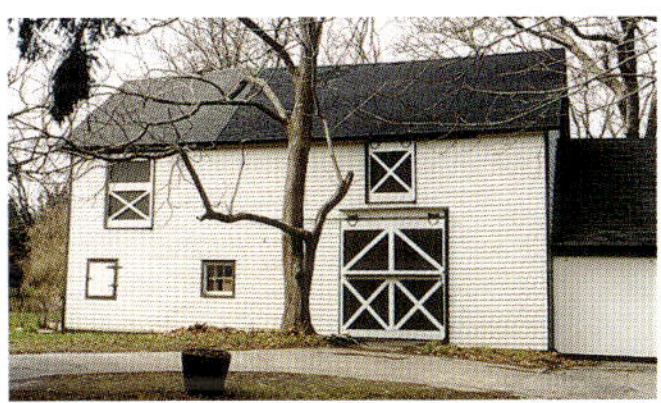

New Suffolk, New Suffolk Road
White clapboard and vertical plank double barn. 117 6 8

New Suffolk, New Suffolk Road
White asbestos-shingled barn with addition. 117 6 30

New Suffolk, New Suffolk Road
White clapboard barn with addition. 117 6 31

New Suffolk, Orchard Street
Gray and white vertical plank barn. 117 5 39

New Suffolk, Orchard Street
Square vertical plank barn. 117 5 40

New Suffolk, Jackson Street
Red vertical plank barn, more than one hundred years old. 117 9 20

New Suffolk, Suffolk Avenue
White vertical plank gambrel-roofed barn. 117 9 11

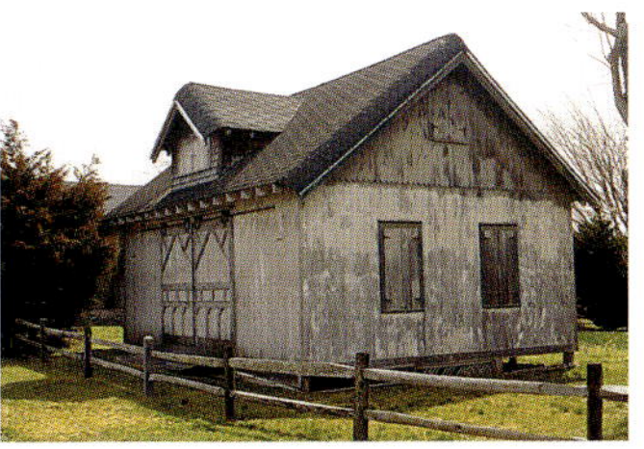

New Suffolk, Jackson Street
Gray vertical plank carriage house for the Lowe estate (1883) on original foundation. 117 10 11

New Suffolk, New Suffolk Avenue
Wood-shingled barn. 116 6 14

New Suffolk, New Suffolk Avenue
Gray vertical plank carriage house; brick ice house. House built on the Stuart Hull Moore farmland estate in 1896. 116 6 10.2

Peconic, Indian Neck Road
Vertical plank English barn with shed addition built in the early twentieth century; three outbuildings. 098 1 27.1

Peconic, Indian Neck Lane
Asbestos-shingled barn with shed addition; carriage house; two outbuildings. 098 1 1.3

Peconic, Leslie's Road
Vertical plank barn with addition, pegged, large loft. 098 1 2.15

Peconic, Indian Neck Lane
Wood-shingled gambrel-roofed barn with flanking wings, bricked cellar, ventilators. The barn has an unusually small gable. It is possible that the barn was divided in half. The house dates to the 1870s and the barn may date to the same time. Second wood-shingled saltbox barn. 086 5 14.2

Peconic, Indian Neck Lane
Vertical plank barn with gray asbestos-shingled shed addition. 086 5 14.4

Peconic, Indian Neck Lane
Gray wood-shingled barn with addition; now a dwelling. 086 4 3

Peconic, Main Road
White stucco-covered double barn built in the nineteenth century, used for tasting room, framing exposed; metal Quonset hut barn; new long barn. 085 2 15

Peconic, Main Road
White board and batten barn built in 1993 with portions of the original barn that it replaced. 085 3 8

Peconic, Main Road
Gray vertical plank gambrel-roofed barn; gray asbestos-shingled barn; three additional buildings including a corncrib. 085 2 16 (pages 152–53)

Peconic, Main Road
White metal pole barn; white vertical plank barn with flanking shed additions, used as a tasting room; white carriage house. 085 2 17.1

Peconic, Main Road
Vertical plank barn; pole barn; metal pole barn. 085 2 18.2 (page 95)

Peconic, Main Road
Beige metal pole barn; cinder-block and vertical plank barn with extensions. 085 2 18.1

Peconic, Main Road
White vertical plank double barn; two additional white outbuildings. Two bays were added to the original three-bay barn for a total of five bays. The older barn's frame is hand hewn, mortise and tenon, the rafters are pinned, and there is no ridge pole. The additional bays have a ridge pole and the rafters are not hand hewn. 085 3 9 (page 19)

Peconic, Main Road
Metal barn; vertical plank barn. 085 3 12.1

Peconic, Main Road
White/yellow vertical plank barn with additions. 085 3 10.3

Peconic, Main Road
White vertical plank double barn; white wood-shingled barn. 085 3 11.4

Peconic, Main Road
Red double metal-covered barn; red metal pole barn; red metal Quonset hut barn. 085 2 9.2

Peconic, Main Road
Yellow vertical plank barn with shed addition. 086 1 15

Peconic, Main Road
Vertical plank double barn; square wood-shingled barn. 086 1 9.1

Peconic, Main Road
Vertical plank double potato barn with addition. 075 1 16

Peconic, Main Road
Vertical plank double barn; small gray vertical plank outbuilding; metal pole barn. 075 1 17.1

Peconic, Main Road
Gray metal pole barn with cinder-block extension built in the early 1990s; green metal pole barn built in 2002. 075 1 20.1

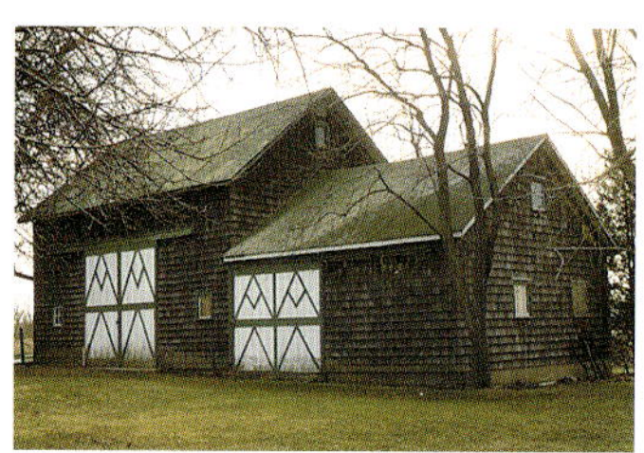

Peconic, Main Road
Wood-shingled double barn; vertical plank square carriage house; additional outbuildings. The large English barn is old. The smaller barn was added in 1927. The old Locust Grove schoolhouse is on the property. 075 6 6.2

Peconic, Main Road
Gray vinyl clapboard double barn. 075 6 7.6

Peconic, Main Road
White vinyl clapboard barn. 075 6 9.5

Peconic, Main Road
Wood-shingled and metal pole double barn; wood-shingled barn. 075 6 11 (pages 72–73)

Peconic, Main Road
Vertical plank gambrel-roofed barn. 075 2 17.2

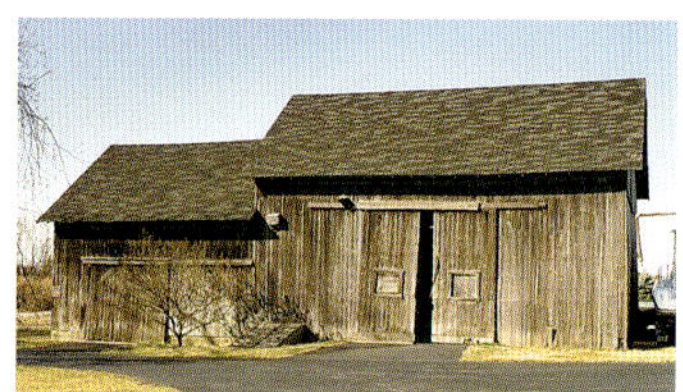

Peconic, Maple Avenue
Vertical plank double barn. Larger barn has English hay barn form. 075 2 9

Peconic, Main Road
White vertical plank double barn; white vinyl-sided barn; white vertical plank carriage house. 075 3 2

Peconic, Main Road
Wood-shingled barn with shed addition; small square vertical plank outbuilding. 069 5 16

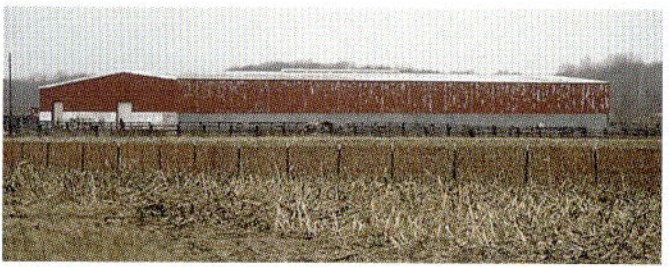

Peconic, Route 48
Vertical plank and wood-shingled triple barn; red cinder-block and metal pole barn. 069 1 7.3

Peconic, Route 48
Wood-shingled barn. 069 1 6

Peconic, Route 48
Wood-shingled barn; green metal pole barn; green asbestos-shingled barn; gray asbestos-shingled barn; greenhouses. 069 4 8.1

Peconic, Route 48
Red vertical plank barn with shed addition. The house was built in 1830 and it is possible that the barn is the same age. 069 1 3 (page 121)

Peconic, Route 48
Beige/brown board and batten gambrel-roofed barn with green-house addition. 068 4 19

Peconic, Route 48
Red vertical plank barn with shed addition. 068 4 20

Peconic, Route 48
Two wood-shingled barns; vertical plank and wood-paneled barn; three additional outbuildings. 068 4 18 (page 17)

Peconic, Route 48
New white vertical plank stable. 074 3 4.1

Peconic, Route 48
Black vertical plank barn built in the early to mid-nineteenth century, original to site; shed now attached to the house is also original. 074 3 11

Peconic, Peconic Lane
Vertical plank barn. 074 3 16

Peconic, Peconic Lane
Two gray board and batten barns. 074 5 8

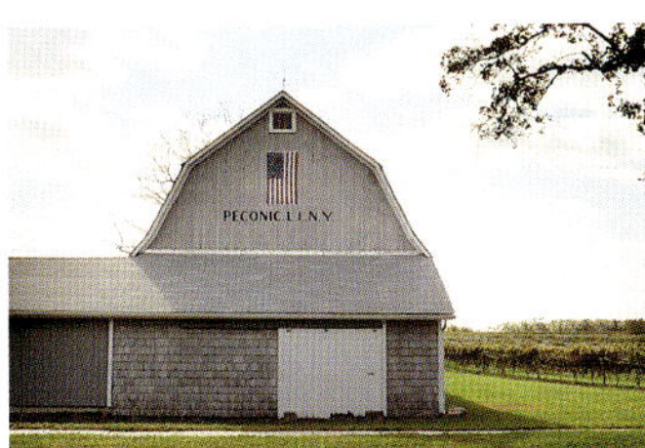

Peconic, Peconic Lane
Gray vertical plank and wood-shingled double barn with addition. 075 5 1

Peconic, Route 48
Beige asbestos-shingled barn with flanking shed additions. The center portion of the structure is a barn that was moved to the site from a farm. 074 4 11

Peconic, Route 48
Two-story stucco structure used in the 1920s by the Atlantic Commission Company for loading potatoes onto railroad cars. 074 4 13

Peconic, Route 48
Gray building that contains a barn moved to the present site. 074 4 5

Peconic, Route 48
Small green asphalt-shingled barn. 074 2 21

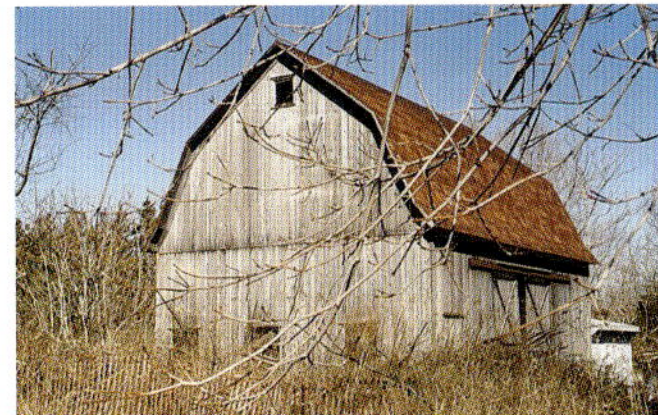

Peconic, Route 48
Vertical plank gambrel-roofed barn. 074 2 37.3

Peconic, Route 48
Gray/green asphalt-shingled barn with shed additions; now a dwelling. 074 1 23

Peconic, Route 48
Red cinder-block and metal Quonset hut barn; four red outbuildings; two corncribs; one circular metal silo. All of the structures were brought to the site from Jericho in the early 1980s. 074 1 36 (page 87)

Peconic, Route 48
Vertical plank barn. This barn was in decline until the land was pre-served. Restoration on the barn was begun in the spring of 2004. 074 4 3.2

Peconic, Route 48
Gray shingled barn; green metal pole barn; vertical plank barn with shed addition. 074 1 37.3 (page 97)

Peconic, Route 48
Large shingled barn with chicken coop addition. 074 1 40.1

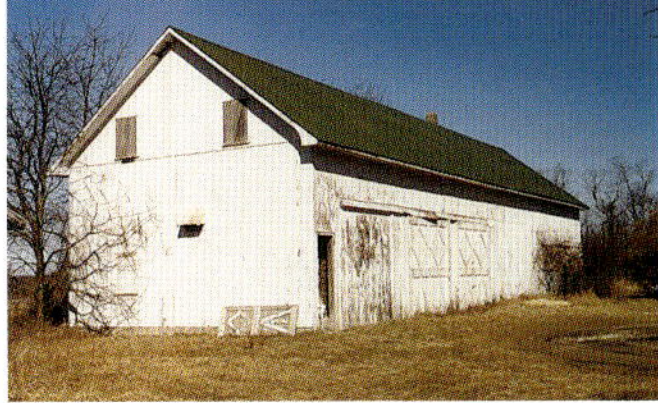

Peconic, Route 48
Square wood-shingled barn; white vertical plank barn. 074 1 42.6

Oregon, Route 48
Deteriorating gambrel-roofed barn. 084 2 4.1

Oregon, Route 48
Stucco buttressed potato barn. 084 2 3.2

Oregon, Route 48
Shingled cinder-block double barn. 084 2 3.3

Oregon, Bridge Lane
Wood-shingled barn with flanking shed additions. 084 2 2.1

Oregon, Bridge Lane
Wood-shingled double English barn with additions; main barn built between 1850 and 1875. 073 2 4

Oregon, Oregon Road
Yellow wood-shingled barn. 084 1 4.4

Oregon, Oregon Road
Gray asphalt-shingled double barn with shed addition; gray vertical plank barn; four outbuildings. 083 2 9.2

Oregon, Oregon Road
Gray shingled and clapboard barn with additions; wood-shingled garage with shed addition. 084 1 11

Oregon, Oregon Road
Beige metal pole barn. 083 3 2.1

Oregon, Oregon Road
Gray vertical plank and wood-shingled double barn with addition; square wood-shingled barn; yellow metal pole barn; gray vertical plank barn; additional outbuildings. 083 2 13.6

Oregon, Oregon Road
Stucco buttressed potato barn. 083 3 5.3 (page 137)

Oregon, Oregon Road
Vertical plank barn. 083 3 5.2

Oregon, Oregon Road
White clapboard and gray asphalt-shingled barn with additions; shed and water tower. 083 2 15

Oregon, Oregon Road
White and gray barn complex (asbestos shingle, vertical plank, and metal sidings); gray asbestos-shingled and white metal barn with addition; gray metal pole barn. 083 2 16

Oregon, Oregon Road
Wood-shingled barn with shed addition. 083 2 17.1

Oregon, Oregon Road
Gray shingled gambrel-roofed double barn with shed addition; wood-shingled barn complex including gambrel-roofed barn; additional outbuildings. 083 1 32.3 (pages 118–19)

Oregon, Depot Lane
Stucco double barn; white wood-shingled outbuilding. 095 2 1.1

Oregon, Oregon Road
Gray asbestos-shingled barn, wood-shingled addition. 095 4 14.3

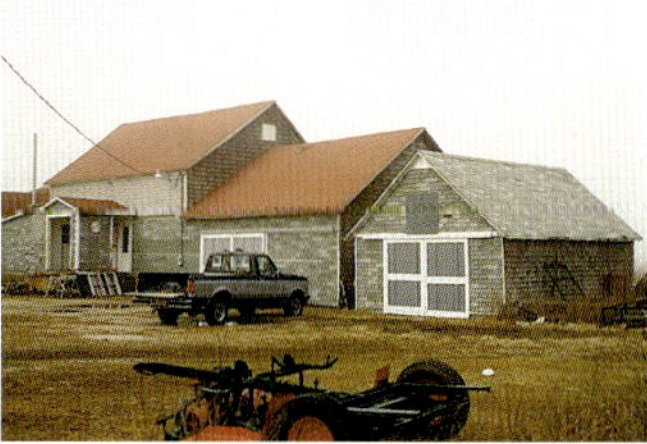

Oregon, Oregon Road
Gray asphalt-shingled double barn with additions; adjacent square barn, same materials. 095 1 11.3

Oregon, Oregon Road
Gray and yellow shingled double barn. 095 1 8.3

Oregon, Oregon Road
Red vertical plank barn; red garage; brick ice house. 095 1 7.1

Oregon, Alvah's Lane
Four connected barns. Materials include vertical plank and beige asbestos shingles. One portion of the quadruple barn is an English hay barn (1850–80), which was moved, attached, and altered for potato storage. 095 4 3.1 (pages 108–09)

Oregon, Alvah's Lane
White asbestos-shingled double barn with shed addition; outbuildings and greenhouses. Main section of the barn is pegged. Portions of the farmhouse are more than two hundred years old. 101 1 19.1

Oregon, Alvah's Lane
Gray vertical plank barn; greenhouses. 101 1 14.3

Oregon, Alvah's Lane
Blue metal pole barn. 095 3 10

Oregon, Oregon Road
Gray asbestos-shingled barn. 095 3 8.1

Oregon, Oregon Road
White vertical plank and asbestos-shingled barn with shed addition; water tower. 095 1 4 (page 141)

Oregon, Oregon Road
Vertical plank barn; wood-shingled barn. 095 1 3.2

Oregon, Oregon Road
New wood-shingled barn on old foundation. 095 1 2

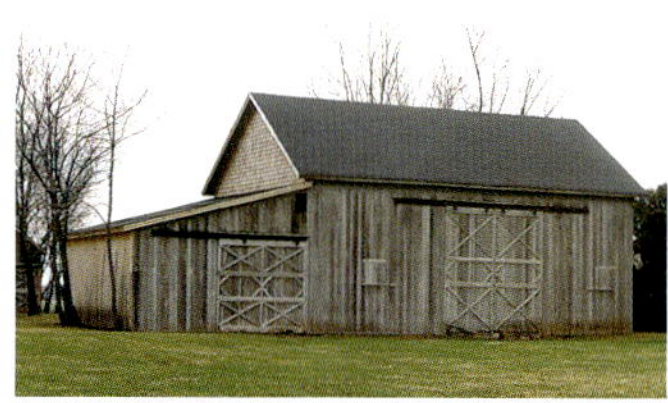

Oregon, Oregon Road
Vertical plank barn with shed addition; vertical plank square barn; vertical plank corncrib. 095 3 3.5

Oregon, Oregon Road
Gray vertical plank triple barn with shed addition (middle barn has a gambrel roof). 095 1 1.2

Oregon, Oregon Road
Gray shingled barn with flanking shed additions; four outbuildings. 094 3 2 (page 45)

Oregon, Oregon Road
Gray vertical plank barn. 100 2 7

Oregon, Oregon Road
Yellow metal pole barn; two gray metal pole barns; five metal silos. 100 2 6.1

Oregon, Oregon Road
Gray shingled barn complex. Main portion of the barn was built in 1917 (noted on rafter). The west barn was moved from next door. 100 4 6.2

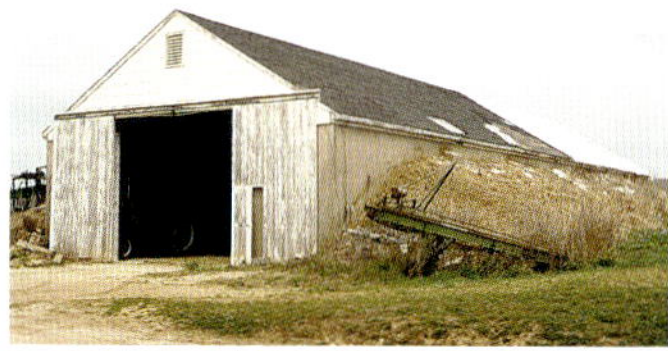

Oregon, Route 48
Vertical plank and shingled barn with additions; banked cinder-block potato barn; numerous sheds. 101 1 4.3

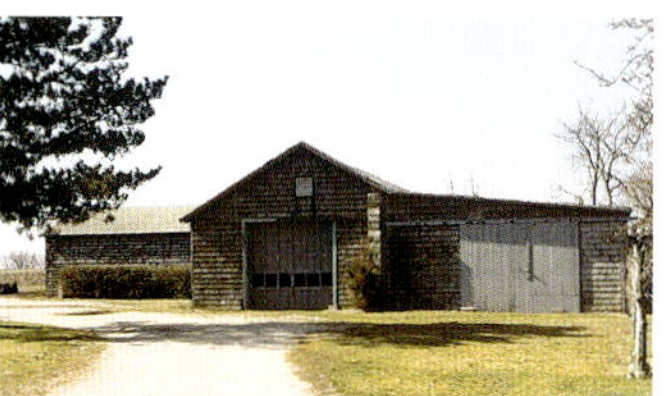

Oregon, Elijah's Lane
Wood-shingled double barn with shed addition; red metal pole barn; chicken coop; two outbuildings. 101 1 2.3

Oregon, Oregon Road
White barn with shed addition; gray cinder-block barn; greenhouses. 100 2 5.2

Oregon, Oregon Road
Gray vertical plank and shingled barn. 100 2 3

Oregon, Oregon Road
Gray vertical plank and wood-shingled barn with shed addition. 100 4 3.2

Oregon, Oregon Road
Gray metal-covered vertical plank barn; two-story double barn; gray asphalt-shingled barn. The two-story double barn was built in 1890 and has been renovated for use as a tasting room. The asphalt-shingled barn was built in 1930–40 and is in agricultural use. 100 4 2.2 (page 25)

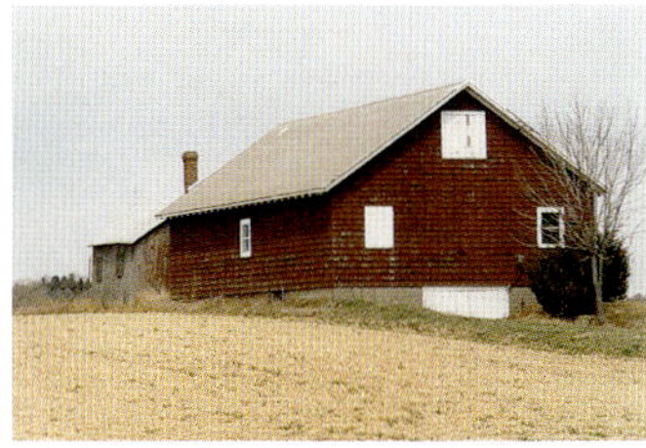

Oregon, Oregon Road
Red wood-shingled barn; outbuilding. 100 2 2

Oregon, Oregon Road
Double English barn; potato barn. 100 2 1

Oregon, Mill Lane
Board and batten–sided barn with flanking shed additions. 100 4 2.1

Oregon, Mill Lane
Vertical plank barn with addition built in 1860. 107 5 2 (page 81)

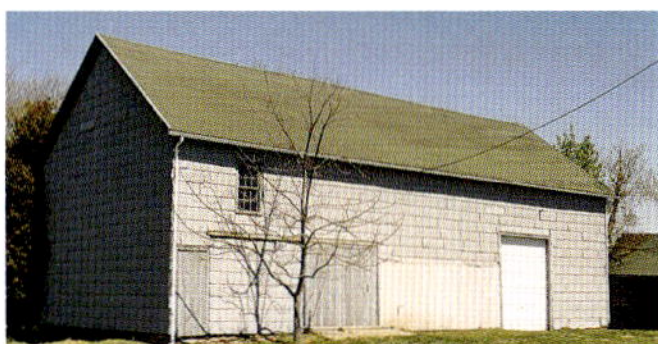

Oregon, Mill Lane
Gray shingled barn, with brick foundation; wood-shingled barn; shed. 107 10 9

Oregon, Wickham Avenue
Gray cinder-block/stucco double barn; metal garage. 107 10 10.1

Oregon, Wickam Avenue
White metal pole barn. 108 1 2

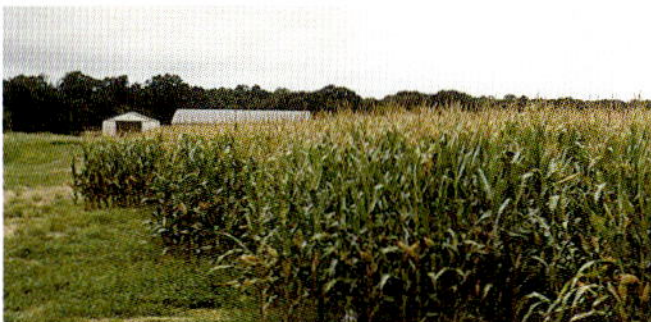

Oregon, Route 48
Gray asphalt-shingled barn; square metal pole barn; additional sheds. 107 10 6

Oregon, Wickham Avenue
Cinder-block potato barn. 107 10 8

Oregon, Mill Lane
White vertical plank barn with shed addition dated 1905; white vertical plank outbuilding. 107 4 3

Oregon, Mill Lane
Stables, white vertical plank. 107 4 2.8

Oregon, Paddock Way
Red wood-shingled and vertical plank barn, gambrel roof, former horse stable. 107 4 2.11

Oregon, Mill Road
White vertical plank barn with shed addition. The house was built in 1865 or earlier. The barn was built circa 1890. An additional building was built as a garage early in the twentieth century. 100 3 16 (page 55)

Oregon, Mill Road
Two cinder-block potato barns. 100 3 12

Oregon, Mill Road
Board and batten yellow barn with shed addition; greenhouses. 107 1 2.1

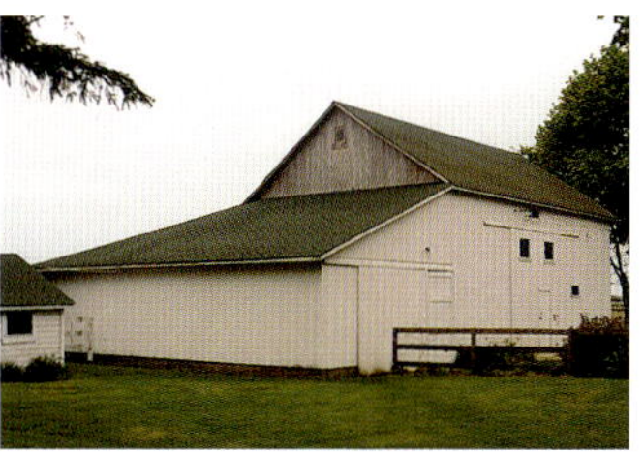

Mattituck, Sound Avenue
White vertical plank barn with shed addition; white shingled carriage house; currently used as an artist's studio. 120 3 1

Mattituck, Sound Avenue
Two beige stucco buttressed attached potato barns built between 1950 and 1955. 120 1 2.1

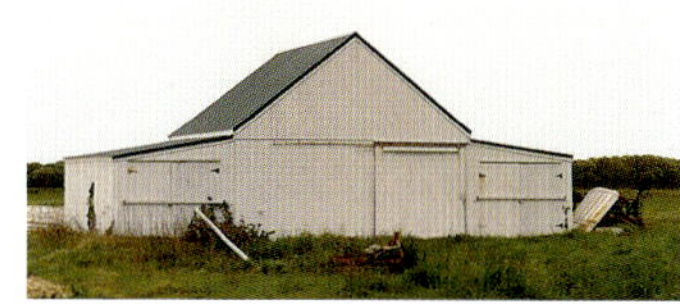

Mattituck, Sound Avenue
Gray shingled barn built in 1968, with wing added in 1977; original barn is a wood-shingled double barn with four addditions; white vertical plank barn with additions; four smaller white vertical plank farm structures, two of which are old; white vertical plank barn with shed additions. 120 1 3, 112 1 7, 120 3 11.10

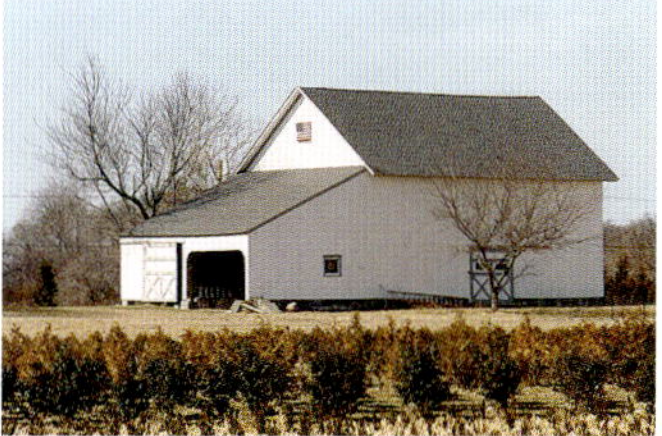

Mattituck, Hallock Lane
White vertical plank barn with shed addition. This was the Hallock family farm. 112 1 8.11

Mattituck, Sound Avenue
Blue wood-shingled barn with shed additions. The original barn (built 1850–60) was lost in the 1938 hurricane. The existing barn was rebuilt over the original stone foundation. The cellar for the barn has double access and is similar in design to an English hay barn. The land was formerly in the Aldrich family. 120 6 1

Mattituck, Sound Avenue
Gray metal pole barn; gray vertical plank barn. 120 3 7

Mattituck, Aldrich Lane
Gray wood- and asbestos-shingled barn with shed additions. 124 1 2.1

Mattituck, Bergen Avenue
White twenty-plus-year-old pole barn; formerly a horse stable now used by vineyard. 120 2 3 1

Mattituck, Sound Avenue
Wood-shingled gambrel-roofed barn with shed addition, built in the late nineteenth century. This area of Mattituck was an important beach landing for unloading liquor during Prohibition. The farmhouse associated with the barn is gone. 120 2 2 (page 125)

Mattituck, Sound Avenue
Vertical plank barn; beige vertical plank/stucco barn. Both barns were recently consructed. 121 1 1.3

Mattituck, Bergen Avenue
White wood-shingled buttressed barn with cupola and shed additions, used during Prohibition for liquor storage. 112 1 16.8

Mattituck, Bergen Avenue
Beige stucco buttressed barn with ventilators and additions, thirty to forty years old; formerly the Bergen farm. 112 1 19

Mattituck, Bergen Avenue
Pink clapboard double barn with shed additions; worker housing. 113 7 19.25

Mattituck, Bergen Avenue
Tan metal pole barn. 113 7 1.1

Mattituck, Bergen Avenue
Gray vertical plank barn, with wood brought out from World's Fair in the 1930s; white clapboard barn; metal pole barn; three gray outbuildings including a corncrib now covered with vertical plank. The farm was established in the 1890s. 113 7 2.5

Mattituck, Bergen Avenue
Gray wood-shingled and vertical plank four-and-a-half- or five-bay barn built in the nineteenth century; two additional gray barns, one vertical plank, one wood shingled; long vertical plank chicken coop/sprout house; small vertical plank former worker cottage. 113 2 26 (page 31)

Mattituck, Sound Avenue
White asbestos-shingled cinder-block potato barn built into the bank; white asbestos-shingled barn, now used as a garage. 121 3 4

Mattituck, Sound Avenue
Long, wood-shingled barn built in the early nineteenth century, now used as both a stable and a dwelling. 121 3 5.3

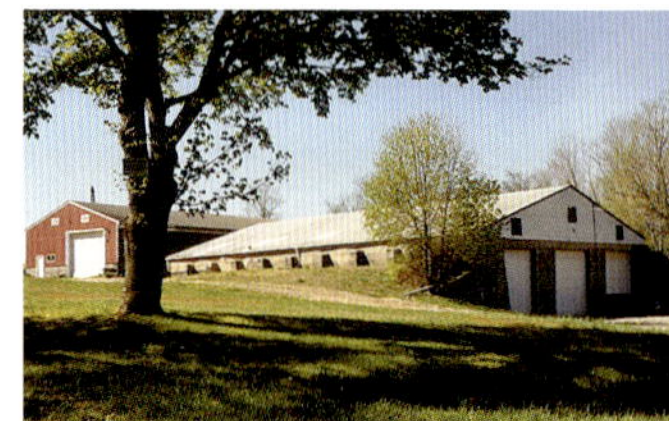

Mattituck, Sound Avenue
Red vertical plank barn on the road was part of the William Unkelbach family dairy farm; beige cinder-block barn was built for potato storage; barns on the hill (two red vertical plank and one blue metal pole) are part of a nursery. 121 3 6

Mattituck, Sound Avenue
Three small beige vertical plank outbuildings and one vertical plank barn, currently used as a stable. 121 1 4.3

Mattituck, Sound Avenue
Square horizontal plank barn with small addition. 121 4 8.1

Mattituck, Cox Neck Road
Two large beige metal plank–covered structures (one a double barn). 121 6 1 (page 93)

Mattituck, Route 48
Small vertical plank barn with wood-shingled roof, not on original foundation; gray metal pole barn in business use. 113 12 10.4

Mattituck, Westphalia Road
Long, narrow blue/gray vertical plank barn, built as a chicken coop, now part of a nursery. 113 13 1.2

Mattituck, Cox Neck Road
Small white vertical plank barn with shed addition. 113 8 1

Mattituck, Luther's Road
Blue/gray clapboard home, which was the carriage house for the adjacent Victorian house. Both were built by the Cox family and sold to the Forman family in 1860. 113 3 7.5

Mattituck, Luther's Road
Vertical plank square barn. 113 3 5

Mattituck, Luther's Road
Double vertical plank barn with large shed addition; smaller vertical plank barn. 106 8 50.4 (pages 36–37)

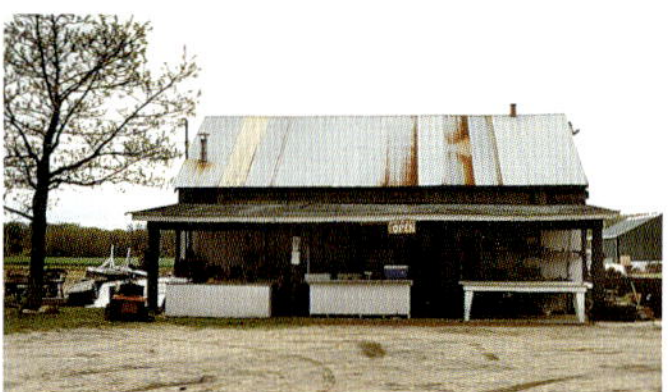

Mattituck, Luther's Road
Square metal pole barn with addition; red vertical plank barn with shed addition; farm stand. 106 9 2.3

Mattituck, Mill Road
White vertical plank and shingled barn with addition, currently used as a dwelling; wood-shingled double barn. 106 9 3

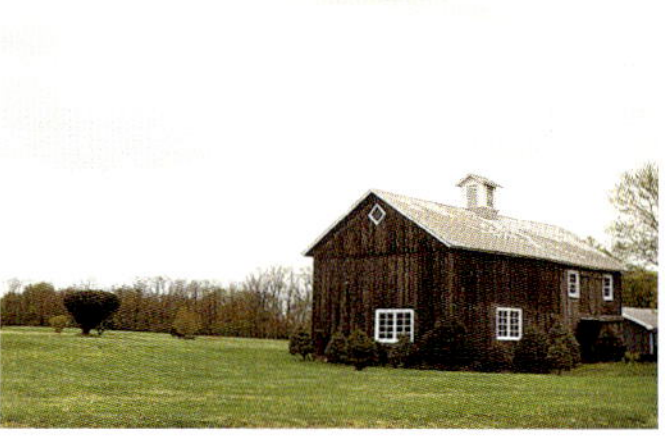

Mattituck, Mill Road
Vertical plank barn built by Richard Cox in 1860–80. The gristmill and house were built in 1821. 106 9 3

Mattituck, Sound Avenue
White wood-shingled barn with second shingled attached barn. 121 5 4.1

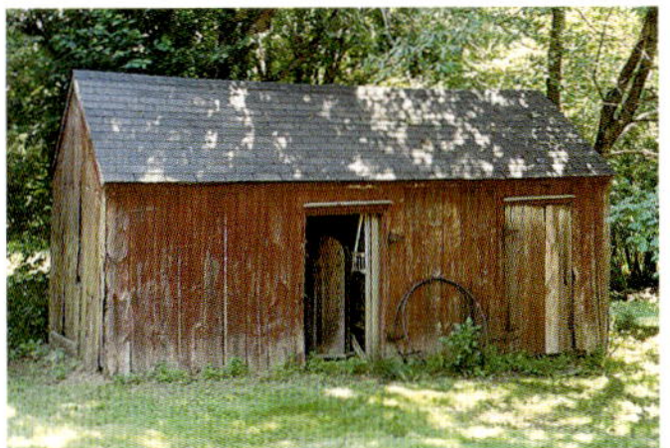

Mattituck, Sound Avenue
Red vertical plank pegged barn with redwood ceiling; new brown vertical plank gambrel-roofed stable; three outbuildings. 122 1 2.2

Mattituck, Sound Avenue
Yellow horizontal plank barn with shed addition. 122 2 8.2

Mattituck, Sound Avenue
Square board and batten barn. 122 2 8.1

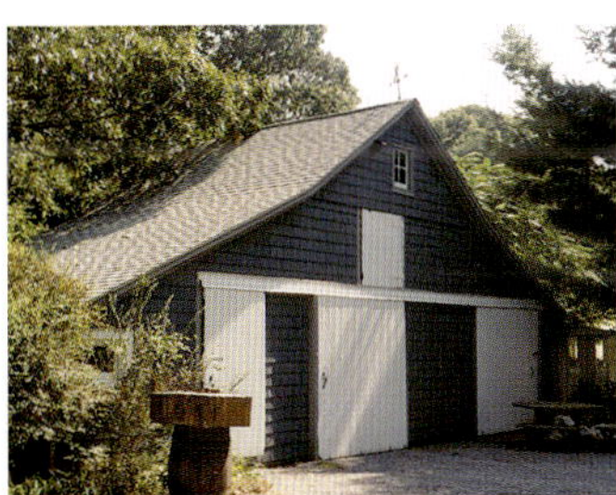

Mattituck, Sound Avenue
Dark gray asphalt-shingled barn with flanking wings. 122 2 11.1

Mattituck, Sound Avenue
Red double barn; barn complex incorporating two barns, one a 50-foot-square gambrel-roofed barn. 141 4 46

Mattituck, Sound Avenue
Blue clapboard double barn attached to a single blue barn. Located on railroad tracks, this was the A&P grading station. 141 3 41

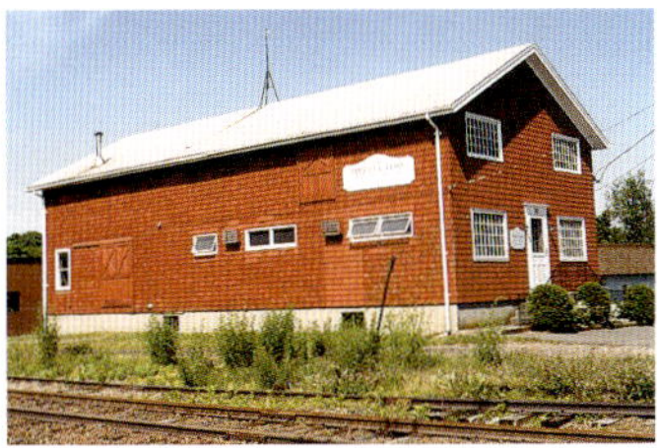

Mattituck, Westphalia Avenue
Red wood-shingled barn. Now an office building, it was once a grading station. 141 3 33

Mattituck, Westphalia Avenue
Vertical plank barn with shed additions; formerly a grading station. 141 3 32.1

Mattituck, Pike Street
Red wood-shingled barn complex consisting of three barns; vertical plank barn with shed addition. These were also grading stations. 141 3 29.1

Mattituck, Love Lane
Small yellow vertical plank barn with shed addition. 140 2 17

Mattituck, Main Road
Two small vertical plank barns, one built in the 1860s by Samuel Sneeden. 140 3 39

Mattituck, Main Road
Vertical plank barn; two smaller vertical plank barns, one red. 140 3 27

Mattituck, Main Road
Vertical plank barn built in the 1870s or 1880s by McGuire and used as a stable and a veterinarian's office. 114 8 6

Mattituck, Mill Lane
Metal pole barn; cinder-block barn; metal Quonset hut barn; English barn built in the 1880s with additions (built in the late nineteenth and early twentieth centuries); cow barn built in the late nineteenth century; chicken coop and corncrib. 115 2 2.3

Mattituck, Main Road
The upper portion of this barn was Mattituck's second schoolhouse, built in 1857. It has been raised one story (brick) and flanked by wood-shingled shed additions. The first Mattituck schoolhouse built before 1829 sits alongside the barn. 115 2 2.1 (page 65)

Mattituck, Main Road
Vertical plank double barn. 115 2 5

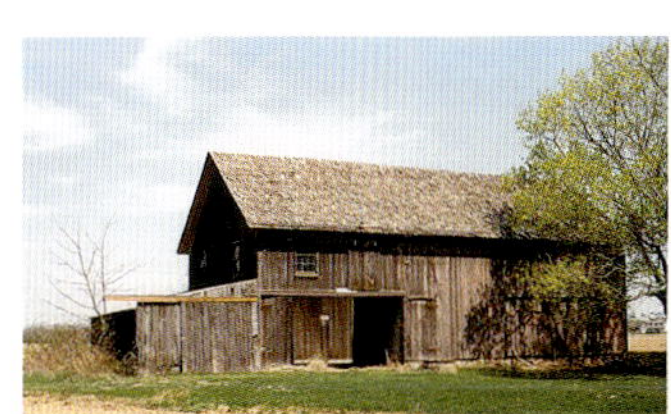

Mattituck, Main Road
Vertical plank barn with shed addition, built in the late nineteenth century. 115 2 6

Mattituck, Main Road
Gray wood-shingled double barn. 115 2 9.1

Mattituck, Main Road
Metal pole barn; white vertical plank and shingled-over cinder-block barn; small vertical plank and asphalt-shingled barn; carriage house/garage. 115 2 10

Mattituck, Main Road
Green asphalt-shingled barn. 108 4 1.1 (page 43)

Mattituck, Rachel's Road
Gray asphalt-shingled barn. One side is full of bees. 108 4 7.44

Mattituck, Elijah's Lane
Gray shingled double barn with flanking wings. 108 4 11.3

Mattituck, Elijah's Lane
Gray vinyl clapboard double barn. 108 3 5.45

Mattituck, Elijah's Lane
Gray metal pole barn; gray vertical plank barn. 108 3 6.2, 108 3 6.3

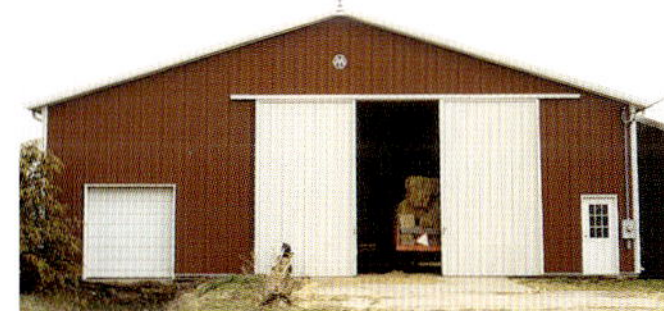

Mattituck, Elijah's Lane
Two red metal pole barns; white summer kitchen and brick milk house. 108 3 1

Mattituck, Route 48
White vertical plank barn with shed addition. 108 2 7

Mattituck, Main Road
Wood-shingled barn with shed addition; outbuilding. 115 8 1

Mattituck, Main Road
Beige seven-bay vertical plank barn with shed addition built in the late nineteenth century. Two additional beige vertical plank outbuildings have timbers from the seventeenth century, including a summer beam. Formerly the Brill farm (1880) and site of Mattituck's first windmill (1660–1710). 115 4 8.5 (page 131)

Mattituck, Suffolk Avenue
White asbestos-shingled barn; green and white metal pole barn. 115 4 33.5

Mattituck, New Suffolk Avenue
Large white vertical plank barn with shed addition; white wood-shingled estate carriage house built in the 1890s by Frank Lupton. Three stables are set deeper in the lot. 115 9 4

Mattituck, New Suffolk Avenue
Yellow vertical plank barn with shed addition. 115 9 3

Mattituck, Reeve Avenue
Gray vertical plank carriage house. 114 11 30

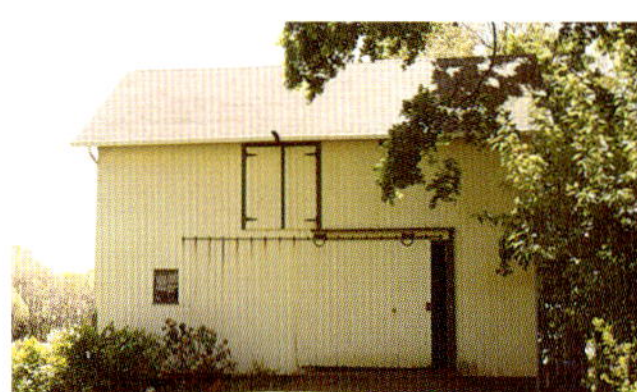

Mattituck, New Suffolk Avenue
Yellow vertical plank barn built by Joseph H. Hudson in the 1880s. 114 12 6

Mattituck, Main Road
New vertical plank and wood-shingled barn built as a stable and dwelling combination. 109 6 9.1

Mattituck, Bay Avenue
Yellowish vertical plank barn with shed additions. 143 3 10

Mattituck, Legion Street
Gray horizontal plank double barn with large shed addition; formerly a produce transfer station. 143 3 10

Mattituck, Old Main Road
Ninety-year-old brick pig barn. 122 7 8.6

Mattituck, Main Road
Red vertical plank barn with shed addition. 122 6 35.4 (page 143)

Mattituck, Peconic Bay Avenue
White vertical plank barn built circa 1910. The house on the bay is thought to have been a summer farmhouse. 128 2 2

Mattituck, Main Road
Gray stucco barn; red vertical plank buttressed pegged barn with shed addition, more than one hundred years old. 125 3 13

Mattituck, Main Road
White vertical plank double barn; white vertical plank barn with addition; large metal pole barn. 125 3 11 (page 71)

Mattituck, Main Road
Gray asphalt-shingled barn with shed addition built in the early nineteenth century, pegged with original wood siding covered by current siding; gray asphalt-shingled potato barn built in the mid-1950s. 125 3 10 (page 149)

Mattituck, Main Road
White vertical plank barn; white wood-shingled barn with shed addition; gray metal barn with addition; large metal pole barn. 125 3 8

Mattituck, Main Road
Wood-shingled barn with shed addition; white vertical plank barn; small cinder-block barn. 125 3 7.2

Mattituck, Main Road
Beige clapboard double barn built in 1997 as a tasting room and for wine storage. 125 1 2.31

Mattituck, Main Road
White shingled gambrel-roofed barn with addition. 125 1 2.8

Mattituck, Main Road
Red vertical plank barn with shed addition. 125 1 2.2

Mattituck, Main Road
Gray shingled barn with shed addition. 125 3 4.1

Mattituck, Franklinville Road
White vertical plank and wood-shingled barn with shed addition. 125 1 3

Mattituck, Franklinville Road
Late eighteenth– or early nineteenth–century English pegged vertical plank double barn with a carriage barn extension; small vertical plank outbuilding. 125 3 1.1

Mattituck, Franklinville Road
Refurbished beige board and batten barn with shed addition. 125 2 1.25

Mattituck, Franklinville Road
Gray metal pole barn. 125 2 2.2

Mattituck, Main Road
Wood-shingled carriage house built in the 1930s or early 1940s. 127 2 8.1

Mattituck, Laurel Lane
Gray metal pole barn. 127 2 2.1

Mattituck, Main Road
Vertical plank barn with shed addition. 125 3 2.3

Mattituck, Main Road
Two vertical plank barns. 127 3 6.1

Mattituck, Peconic Bay Boulevard
Red wood-shingled carriage house; now used as a dwelling; formerly the Baum estate. 145 2 11